The Digital Interface Handbook

The Digital Interface Handbook

Second edition

Francis Rumsey
and
John Watkinson

Focal Press
An imprint Butterworth-Heinemann Ltd
Linacre House, Jordan Hill, Oxford OX2 8DP

 A member of the Reed Elsevier plc group

OXFORD   LONDON   BOSTON
MUNICH   NEW DELHI   SINGAPORE   SYDNEY
TOKYO   TORONTO   WELLINGTON

First published 1993
Reprinted 1994
Second edition 1995
Reprinted 1995

British Library Cataloguing in Publication Data
A catalogue record for this book is available from the British Library

Library of Congress Cataloguing in Publication Data
A catalogue record for this book is available from the Library of Congress

ISBN 0 240 51396 7

Printed and bound in Great Britain by Clays Ltd, St Ives plc

Contents

Chapter 6 Synchronization in digital audio interfacing   148

Chapter 7 Practical audio interfacing   168

Chapter 8 Digital video interfaces   185

Preface

Audio and video systems are increasingly digital rather than analog. The trend towards all-digital installations in broadcasting, recording and post-production requires that not only is the equipment digital but that the interconnections between devices are digital, since this is the only way to ensure that quality is maintained when signals are copied or transferred. In a large installation a signal may pass through many pieces of equipment, and if analog interconnects were used this would require the signal to suffer repeated stages of analog-to-digital and digital-to-analog conversion, with a consequent degrading effect on signal quality. Digital interconnects in conjunction with digital devices allow an unlimited number of stages of copying or transfer with no degradation of programme quality due to the interconnect. A signal passed over a digital interconnect also has the advantages of immunity to crosstalk, interference and noise, which is especially important when many signals share the same trunking.

In this book we have brought digital audio and digital video together, firstly because once converted they are both data signals and similar criteria apply to the handling of each; and secondly because today's engineers in the television, multimedia and music fields are increasingly having to understand both sound and pictures. We have covered the subject of digital interfacing from the bottom up, tackling both 'analog' problems (such as electrical matching, cabling, transmission lines) and 'digital' matters (such as channel coding and data separation), since it is vital to understand that conventional circuit theory remains important when considering digital communications. In this book the reader will find fundamental theory, an introduction to digital audio and video, commentary on all the commonly used interfaces, and guidance on 'real-world' implementations and systems. The information contained in this book has been gleaned from a wide range of international sources, and contains data which is often very difficult to discover without spending months collecting miscellaneous documents. We have attempted to sift through the morass of standards and to highlight the important differences between them, showing how incompatibilities may arise which prevent correct communication between systems. The information is as up to date as we could make it in such a rapidly evolving field. We have also incorporated a great deal of practical experience with digital interfaces, which is an important feature since devices do not always behave according to the standards. As such, the book should prove a useful aid when troubleshooting all-digital installations which refuse to work properly.

A great deal more information will be found in this second edition on the subject of digital video. Now that the subject is reaching a greater degree of standardization, and equipment is entering service in greater quantities, it is possible to include more on the practical applications of video, including reliability testing techniques, EDH, and signature analysis. There are numerous subtle differences between the interfaces, and in audio there is also the question of consumer versus professional implementation which causes some difficulties. Further, although the issue of synchronization has been inherent in video ever since it started, synchronization is a relatively new subject in audio – having been largely unnecessary in analog installations, but being closely involved with digital audio systems. For this reason a whole chapter is devoted to synchronization in digital audio, with discussion of the timing relationships between digital audio and video.

It perhaps goes without saying that this book is not a standards document and makes no pretensions to being one. The place of standards documents is to lay out the *status quo* in formal, often legalistic, language, whereas our book should be viewed as a means of shedding light on the standards – showing how they have been implemented in the real world, and how they can be made to work properly. Ideally one should have both, and readers are advised to use this book in conjunction with the relevant standards.

Our thanks are due to all those who helped us in putting this book together, particularly to Julian Dunn of Prism Sound, Robin Caine and Graham Roe of Pro-Bel, and Jim Wilkinson of Sony Broadcast and Professional. Without the Audio Engineering Society as an international forum for the exchange of ideas much of this book would not have been possible.

Francis Rumsey
John Watkinson

January 1995

Chapter 1

Introduction to interfacing

1.1 The need for digital interfaces

1.1.1 Transparent links

Digital audio and video systems make it possible for the user to maintain a high and consistent sound or picture quality from beginning to end of a production. Unlike analog systems, the quality of the signal in a digital system need not be affected by the normal processes of recording, transmission or transferral over interconnects, but this is only true provided that the signal remains in the digital domain throughout the signal chain. Converting the signal to and from the analog domain at any point has the effect of introducing additional noise and distortion, which will appear as audible or visual artefacts in the programme material. Herein lies the reason for adopting digital interfaces when transferring signals between digital devices – it is the means of ensuring that the signal is carried 'transparently', without the need to introduce a stage of analog conversion. It enables the receiving device to make a 'cloned' copy of the original data, which may be identical numerically and temporally.

1.1.2 The need for standards

The digital interface between two or more devices in an audio or video system is the point at which data is transferred. Digital interconnects allow programme data to be exchanged, and they may also provide a certain capacity for additional information such as 'housekeeping data' (to inform a receiver of the characteristics of the programme signal, for example), text data, subcode data replayed from a tape or disk, user data, communications channels (e.g. low quality speech) and perhaps timecode. These applications are all covered in detail in the course of this book. Standards have been developed in an attempt to ensure that the format of the data adheres to a convention and that the meaning of different bits is clear, in order that devices may communicate correctly, but there is more than one standard and thus not all devices will communicate with each other. Furthermore, even between devices using ostensibly the same interface there are often problems in communication due to differences in the level or completeness of implementation of the standard, the effects of which are numerous. Older devices may have problems with data from newer devices or vice versa, since the standard may have been modified or clarified over the years.

As digital audio and video systems become more mature the importance of correct communication also increases, and the evidence is that manufacturers are now beginning to take correct implementation of interface standards more seriously, since more people are adopting fully digital signal chains. But it must be said that at the same time the applications of such technology are becoming increasingly complicated, and the additional data which accompanies programme data on many interfaces is becoming more comprehensive – there being a wide range of different uses for such data. Thus manufacturers must decide the extent to which they implement optional features, and what to do with the data which is not required or understood by a particular device. Eventually it is likely that digital interface receivers will become more 'intelligent', such that they may analyse the incoming data and adapt so as to accommodate it with the minimum of problems, but this is rare in today's systems and would currently add considerably to the cost of them.

1.1.3 Digital interfaces and programme quality

To say that signal quality cannot be affected provided that the signal remains in the digital domain is a bold statement and requires some qualification, since there will be cases where signal processing in the digital chain may affect quality. The statement is true if it is possible to assume that the sampling rate of the signal, its resolution (number of bits per sample) and the method of quantization all remain unchanged (see Chapter 2). Further one must assume that the signal has not been subjected to any processing, since filtering, gain changing, and other such operations may introduce audible or visual side effects. Operations such as sampling frequency conversion and changes in resolution (say, between 20 and 16 bits per sample) may also introduce artefacts, since these can never be 'perfect' processes. Therefore an operation such as copying a signal digitally between two recording systems with the same characteristics is a transparent process, resulting in absolutely no loss of quality (see Figure 1.1), but copying between two recorders with different sampling

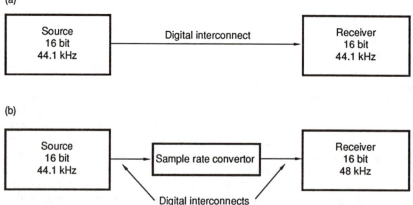

Figure 1.1 (a) A 'clone' copy may be made using a digital interconnect between two devices operating at the same sampling rate and resolution. (b) When sampling parameters differ, digital interconnects may still be used, such as in this example, but the copy will not be a true 'clone' of the original

rates via a sampling frequency convertor is not (although the side effects in most cases are very small).

Confusion arises when users witness a change in sound or picture quality even in the former of the above two cases, leading them to suggest that digital copying is *not* a transparent process, but the root of this problem is not in the digital copying process – it is in the digital-to-analog conversion process of the device which the operator is using to monitor the signal, as discussed in greater detail in Chapter 2. It is true that timing instabilities and (occasionally) errors may arise when signals are transferred digitally between devices, but both of these are normally correctable or avoidable within the digital domain. Since data errors are extremely rare in digitally interfaced systems, it is timing instabilities which will have the most likely effect on the convertor. Poor quality clock recovery in the receiver and lack of timebase correction in the convertor often allow timing instabilities resulting from the digital interface to affect programme quality when it is monitored in the analog domain. This does *not* mean that the digital programme itself is of poor quality, simply that the convertor is incapable of rejecting the instability. Although it is difficult to ensure low jitter clock recovery in receivers, especially with the stability required for very high convertor resolutions (e.g. 20 bits in audio), it is definitely here that the root of the problem lies and not really in the nature of digital signals themselves, since it is possible to correct such instabilities with digital signals but not normally possible with analog signals. This is discussed further in Chapter 2 and in section 6.4.2, and for additional coverage of these topics the reader is referred to Rumsey[1] and Watkinson[2,3].

1.2 Analog and digital communication compared

Before going on to examine specific digital audio and video interfaces it would be useful briefly to compare analog and digital interfaces in general, and then to look at the basic principles of digital communication.

In an analog wire link between two devices the baseband audio or video signal from a transmitter is carried directly in the form of variations in electrical voltage. Any unwanted signals induced in the wire, such as radio frequency interference (RFI) or mains hum will be indistinguishable from the wanted signal at the receiver, as will any noise or timing instability introduced between transmitter and receiver, and these are likely to affect signal quality (see Figure 1.2). Techniques such as balancing (see section 1.6.1) and the use of transmission lines (see section 1.6.3) are used in analog communication to minimize the effects of long lines and interference. Forms of modulation are used in analog links, especially in radio frequency

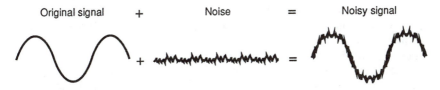

Original signal   +   Noise   =   Noisy signal

Figure 1.2 If noise is added to an analog signal the result is a noisy signal. In other words, the noise has become a feature of the signal which may not be separated from it

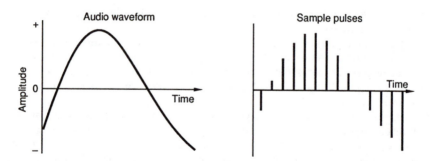

Figure 1.3 In pulse amplitude modulation (PAM) a regular chain of pulses is amplitude modulated by the baseband signal

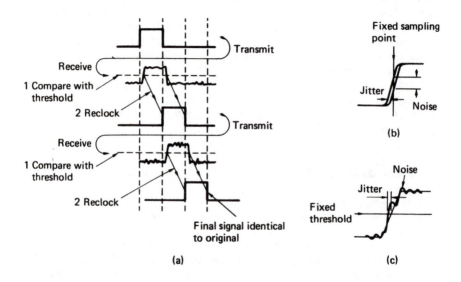

Figure 1.4 (a) A binary signal is compared with a threshold and re-clocked on receipt, thus the meaning will be unchanged. (b) Jitter on a signal can appear as noise with respect to fixed timing. (c) Noise on a signal can appear as jitter when compared with a fixed threshold

transmission, whereby the baseband (unmodulated) signal is used to alter the characteristics of a high frequency carrier, resulting in a spectrum with a sideband structure around the carrier. Modulation may give the signal increased immunity to certain types of interference[4], but modulated analog signals are rarely carried over wire links within a studio installation.

Pulse amplitude modulation (PAM) is a means of modulating an analog signal onto a series of pulses, such that the amplitude of the pulses varies according to the instantaneous amplitude of the analog signal (see Figure 1.3) and this is the basis of pulse code modulation (PCM) whereby the amplitude of PAM pulses is quantized, resulting in a binary signal that is exceptionally immune to noise and interference since it only has two states. This process normally takes place in the analog-to-digital (A/D) convertor of an audio or video system, described in more detail in Chapter 2. PCM is the basis of all current digital audio and video interfaces, and, as shown in Figure 1.4, the effects of unwanted noise and timing instability may be rejected by re-clocking the binary signal and comparing it with a fixed threshold. Provided that the receiving device is able to distinguish between the two states, one and zero, and can determine the timing slot in which each binary digit (bit) resides, the system may be shown to be able to reject any adverse effects of the link. Over long links a digital signal may be reconstructed regularly, in order to prevent it from becoming impossibly distorted and difficult to decode. Of course there will be cases in which an interfering signal will cause a sufficiently large effect on the digital signal to prevent it from being correctly reconstructed, but below the threshold at which this happens there will be no effect at all.

Compared with an analog baseband signal, its digital counterpart normally requires a considerably greater bandwidth, as can be seen from the diagrams above, but the advantage of a digital signal is that it can normally survive over a channel with a relatively low signal-to-noise ratio. Another advantage of digital communications is that a number of signals of different types may be carried over the same physical link without interfering with each other, since they may be time-division multiplexed (TDM), as discussed in section 1.5.5.

1.3 Quantization, binary data and wordlengths

When a PAM pulse is quantized it is converted into a PCM word of a certain length or resolution. The number of bits in the word determines the accuracy with which the original analog signal may be represented – a larger number of bits allowing more accurate quantization, resulting in lower noise and distortion. This process is covered further in Chapter 2, and thus will not be discussed further here; suffice it to say that for digital audio word lengths of up to 24 bits may be considered necessary for high sound quality (with 16–20 bits being typical in modern systems), whereas for digital video a smaller number of bits per sample (typically 8–10) is adequate. Since the number of bits per sample is related directly to the signal-to-noise (S/N) ratio of the system it can be deduced that the S/N ratio required for high quality sound is greater than that required for high quality pictures.

Although the resolution of audio samples may be greater than that of video samples, the sampling rate of a video system is much higher than that required for audio. Thus the total amount of data required per second to represent a moving video picture is considerably greater than that required to represent a sound signal. (In all cases it is assumed that no form of data reduction is used.) This has important implications when considering the requirements for different types of digital interface.

In computer terminology, 8 bits is a *byte*, and this is the unit of storage often used in computer systems, even though data may actually be handled in wordlengths considerably longer than 8 bits. A binary *word* is sometimes confused with a *byte*, but a word can be of virtually any length, whereas a byte may only ever be 8 bits. The bit with the greatest 'weight' in a binary word (the leftmost, or the highest power of two, when written down in conventional form) is called the *most significant bit*, and the bit with the least weight ($2^0=1$, which is normally the rightmost bit) is called the *least significant bit*. The following example illustrates the point:

Take the 8 bit binary value '01011101':

	MSB							LSB
Binary	0	1	0	1	1	1	0	1
Binary weight	2^7	2^6	2^5	2^4	2^3	2^2	2^1	2^0
Decimal weight	128	64	32	16	8	4	2	1
Decimal equivalent	0 +	64 +	0 +	16 +	8 +	4 +	0 +	1 = 93

A kilobit (kb) is 1024 bits (2^{10} bits), and a kilobyte (kbyte) is 1024 bytes. A megabit (Mb) is 1024 kilobits. Confusingly, in communications terminology a data rate of a kilobit per second represents 1000 bits per second, not 1024 bits per second.

1.4 Serial and parallel communications

The bits of a binary word may be transmitted either in parallel or serial form (see Figure 1.5). In the parallel form each bit is carried over a separate communications channel and the result is at least as many channels as there are bits in the word (there are normally additional lines for controlling the exchange of data on a parallel interface). Thus a 24 bit parallel interface would require at least 24 wires, an earth return, a clock line, and a number of address and handshaking lines. Such an approach is normally used for short distance communications ('buses') within a digital device, but is bulky and uneconomical for use over longer distances.

When data is carried serially it only requires a single channel (although electrically that channel may consist of more than one wire), and this makes it economical and simple to implement over large distances. On a serial interface the bits of a word are sent one after the other, and thus it tends to be slower than the parallel equivalent, but the two are so different that to say this is really oversimplifying the matter since there are some extremely fast serial interfaces around. Some serial interfaces carry clock and control information over the same channel as the data, whereas others accompany the data channel with a number of additional parallel lines and a clock signal to control the flow of data between devices. Depending on the standard protocol in use it is possible for serial data to be sent either MSB first or LSB first (see section 1.3), and knowing the convention is clearly important when interpreting received data. These matters are examined further in the next section.

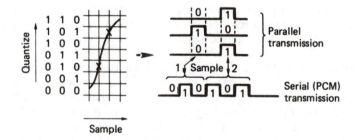

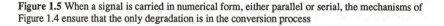

Figure 1.5 When a signal is carried in numerical form, either parallel or serial, the mechanisms of Figure 1.4 ensure that the only degradation is in the conversion process

1.5 Introduction to interface terminology

It is important to understand certain fundamental principles which will arise regularly in the discussion of interfacing and communications.

1.5.1 Data rate versus baud rate

The data rate of an interface is the rate at which information is carried (sometimes referred to as the information rate), whereas the baud rate of an interface is the *modulation* rate or number of data 'symbols' per second. Although in many cases the two are equivalent, modulation schemes exist which allow for more than 1 bit to be carried per baud, such as by using multi-level encoding, by using a form of phase-shift keying or other such channel code (see section 3.3).

Data rate is normally quoted as so many kilo- or megabits per second (kb/s, Mb/s) and this must normally include any capacity for control and additional data. The term *baud rate* is used more widely in computer and telecommunications systems than it is in audio and video interfacing, but it is useful to understand the distinction.

1.5.2 Synchronous, asynchronous and isochronous communications

The receiving device must be able to determine in which time slot it is to register each bit of data which arrives and there are two approaches to achieving this end. In synchronous communications a clock signal normally accompanies the data, either on a separate wire or modulated with the data (see Chapter 3) and this is used to synchronize the receiver's clock to that of the transmitter. Each bit of data may be latched at the receiver on one of the edges of a separate clock, or the clock may be extracted from the modulated data using a suitable phase-locked loop (PLL), as described in section 3.2.4.

In asynchronous communications the clocks of the transmitter and receiver are not locked directly, but must have an almost identical frequency. The tolerance is often around ±1%. In such a protocol each byte of data is prefixed with a start bit and followed by one or more stop bits (see Figure 1.6) and the phase of the receiver's

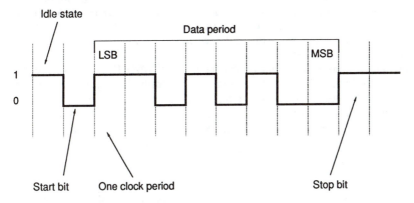

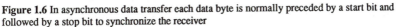

Figure 1.6 In asynchronous data transfer each data byte is normally preceded by a start bit and followed by a stop bit to synchronize the receiver

clock is adjusted at the trailing edge of the start bit. The following data bits are then clocked in according to the receiver's clock, which should remain sufficiently in phase with the transmitted data over the duration of 1 byte to ensure correct reception. The receiver's clock is then resynchronized at the start of the next data byte. Such an approach is often used in computer systems for exchange of data with remote locations over a modem, for example, where the gaps between received bytes may be variable and data flow may not be regular.

In isochronous systems one master clock is used to synchronize all receiving devices, and data transmission and reception is clocked with relation to this source.

1.5.3 Uni- and bidirectional interfaces

In a unidirectional interface data may only be transmitted in one direction, and no return path is allowed from the receiver back to the transmitter. In a bidirectional interface a return path is provided and this allows for two-way communications. The return path is often used in a simple serial situation to send back handshaking information to the transmitter, telling it whether or not the data was received satisfactorily. Unfortunately such an approach is not particularly useful in digital audio and video interfacing because the data in these situations is transferred in real time, making re-transmission in the case of an error an unrealistic concept.

A simplex interface is one which operates in one direction only; a half-duplex interface is one which operates in both directions, but only one at a time; and a full-duplex interface is one capable of simultaneous transmission and reception.

1.5.4 Clock signals

A clock signal may be needed, as described above, to synchronize the receiver. In digital audio and video interfacing a variety of methods are used to ensure synchronism between transmitter and receiver which are discussed in detail in the sections concerned. In general two important clock frequencies exist, that is the 'word clock' which indicates the sampling frequency or rate of sample words over the interface, and the 'bit clock' which indicates the rate of individual data bits. Some interfaces, such as the AES/EBU audio interface, combine the bit clock with the data using a modulation scheme or channel code known as bi-phase mark and indicate the starts of sample words using a violation of the modulation scheme which is easily detected by the receiver. This approach avoids the need for additional lines to carry clock signals and the data is said to be 'self-clocking'. In contrast, an interface such as Mitsubishi's audio interface carries the bit clock and word clock signals on individual wires.

An alternative approach is that used in the so-called 'MADI' audio interface (see section 4.8) in which the transmitter and receiver are both locked to a common reference clock signal and the data is transmitted asynchronously, using a buffer at both ends of the interface to allow for flexibility in timing. This is a form of isochronous approach. The different methods are summarized in Figure 1.7.

1.5.5 Multiplexing

A multiplexed interface is one which carries more than one data signal over a single channel. This is normally achieved using time-division multiplexing (TDM), whereby different time slots in the data stream carry different kinds of information

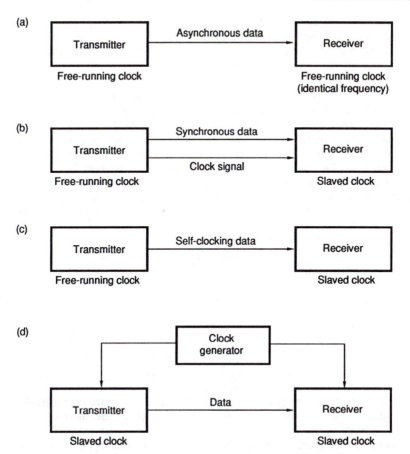

Figure 1.7 A number of approaches may be used to ensure synchronization in data transfer.
(a) Asynchronous transfer relies on transmitter and receiver having identical frequencies, requiring
only the receiver's clock phase to be adjusted by the start bit of each byte. (b) In synchronous
transfer the data signal is accompanied by a separate clock. (c) A form of synchronous transfer
involves modulating the data in such a way that a clock signal is part of the channel code. (d) In the
isochronous approach both devices are locked to a common clock

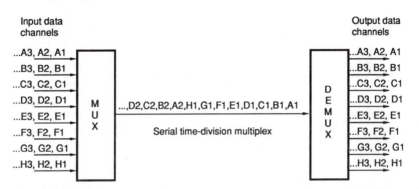

Figure 1.8 In a serial time-division multiplex a number of input channels share a single high-speed
link. Each time slot carries a data packet for each channel

(see Figure 1.8). Here the channel capacity is divided up between the data streams using it, and a multiplexer takes control of the insertion of data into the correct time slots. A demultiplexer at the receiving end extracts the data from each time slot and produces a number of separate data streams.

The AES/EBU audio interface is a simple example of a multiplexed interface, since left and right channel data are multiplexed over the same communications channel, with samples of data for each channel taking alternate time slots on the interface. The MADI interface is a more complicated example in which 56 channels of audio data are multiplexed onto one communications channel – a result of which is that the data rate of the communications channel must be extremely high in order to ensure that data for each audio channel can still be carried in real time.

Thus a multiplexed interface must have a data rate high enough to handle the total data rate of all the data streams which are to be multiplexed onto it, otherwise delays will arise as data queues to use the link. In a computer network it is often acceptable to have short delays where the network is shared between many users, since one can wait for a file to load from a remote server, but for real-time applications such as audio and video it is clearly unacceptable to have a shared network which cannot always carry each channel without breaks or queuing.

1.5.6 Buffering

Buffering is sometimes used at transmitting and receiving ends of an interface to store a number of samples of data temporarily. The buffer is normally a RAM (Random Access Memory) store configured in the FIFO (First In, First Out) mode whose input may be addressed separately to its output (see Figure 1.9). Using such a buffer it is possible to iron out irregularities in data flow, since erratic data arriving at the input may be stored in the buffer and read out at a more regular rate after a short delay. The approach will be successful provided that the average rate of flow into the buffer equals the average rate of flow out of it, and the buffer is big enough to accommodate the irregularities which may arise at its input, otherwise the buffer will either overflow or become empty after a time. A disadvantage of the approach is the delay which arises between input and output, which may be undesirable.

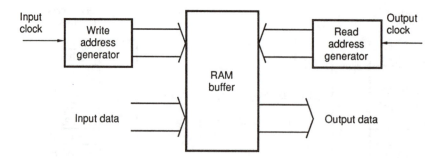

Figure 1.9 A memory buffer may be used as a temporary store to handle irregularities in data flow. Data samples are written to successive memory locations and read out a short time later under control of the read clock

1.6 The electrical interface

Although digital interfaces are primarily concerned with the transfer of binary data, there are 'analog' problems to be considered, since the electrical characteristics of the interface such as the type of cable used, its frequency response and impedance will affect the ability of the interface to carry data signals over distances without distortion.

1.6.1 Balanced and unbalanced compared

In an unbalanced interface there is one signal wire and a ground, and the data signal alternates between a positive and negative voltage with respect to ground (see Figure 1.10). The shield of the cable is normally connected to the ground at the transmitter end, and may or may not be connected at the receiver end depending on whether there is a problem with earth loops (a situation in which the earths of the two devices are at different potentials, causing a current to circulate between them, sometimes resulting in hum induction into the signal wire). The unbalanced interface tends to be quite susceptible to interference, since any unwanted signal induced in the data wire will be inseparable from the wanted signal.

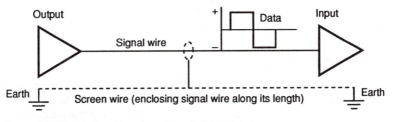

Figure 1.10 Electrical configuration of an unbalanced interface

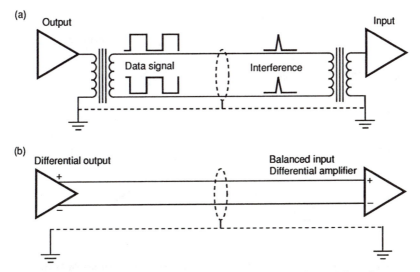

Figure 1.11 Electrical configuration of a balanced interface. (a) Transformer balanced. (b) Electronically balanced

In a balanced interface there are two signal wires and a ground (see Figure 1.11) and the interface is terminated at both ends either in a differential amplifier or a transformer. The driver drives the two legs of the line in opposite phase, and the advantage of the balanced interface is that any interfering signal is induced equally into the two legs, *in phase*. At the receiver any so-called 'common mode' signals are cancelled out either in the transformer or differential amplifier, since such devices are only interested in the *difference* between the two legs. The degree to which the receiver can reject common mode signals is called the common mode rejection ratio (CMRR). Although it is a generally held belief that transformers offer better CMR than electronically balanced lines, the performance of modern differential line drivers and receivers is often as good. The advantage of transformers is that they make the line truly 'floating', that is independent of the ground, and there is no DC coupling across them, thus isolating the two devices.

The balanced interface therefore requires one more wire than the unbalanced interface, and will usually have lines labelled 'Ground', 'Data+' and 'Data–', whereas the unbalanced interface will simply have 'Ground' and 'Data'. For temporary test set-ups it is sometimes possible to interconnect between balanced and unbalanced electrical interfaces or vice versa, by connecting the unbalanced interface between the two legs of the balanced one, or between the ground and one leg, but often the voltages involved are different, and one must take care to ensure that the two data streams are compatible. Balancing transformers are available which will convert a signal from one form to the other.

1.6.2 Electrical interface standards

The different interface standards specify various peak-to-peak voltages for the data signal, and also specify a minimum acceptable voltage at the receiver to ensure correct decoding (this is necessary because the signal may have been attenuated after passing over a length of cable). Quite commonly serial interfaces conform to one of the international standard conventions which describe the electro-mechanical

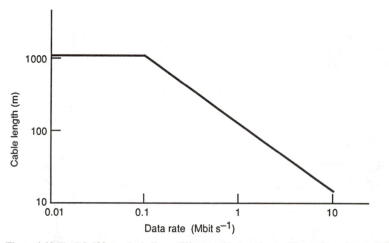

Figure 1.12 The RS-422 standard allows different cable lengths at different frequencies. Shown here is the guideline for data signalling rate versus cable length when using twisted-pair cable with a wire diameter of 0.51 mm. Longer distances may be achieved using thinner wire

characteristics of data interfaces, such as RS-422[5] which is a standard for balanced communication over long lines devised by the EIA (Electronics Industries Association). For example, the AES3-1992 audio interface is designed to be able to use RS-422 drivers and receivers and specifies a peak-to-peak voltage between 2 and 7 volts at the transmitter (when measured across a 110 ohm resistor with no interconnecting cable present), and also specifies a minimum 'eye-height' (see section 3.2.4) at the receiver of 200 mV.

(In passing it should be noted that standards such as RS-422 are mostly only electrical or electro-mechanical standards, and do not say anything about the format or protocol of the data to be carried over them.)

Unbalanced serial interfaces such as RS-232 are not used in audio and video systems, since they are not designed for the high data rates and long distances involved, and are more suited to telecommunications. The voltages involved in an RS-232 interface can be up to 25 V, and thus it is not recommended that one should interconnect an RS-232 output with an RS-422 input (which may be damaged by anything above around 12 volts). An RS-422 interface is designed to carry data at rates up to 100 kbaud over a distance of 1200 metres. Above this rate the distance which may be covered satisfactorily drops with increasing baud rate, as shown in Figure 1.12, depending on whether the line is terminated or not (see section 1.6.3).

Other manufacturer-specific interfaces often use TTL levels of 0–5 volts over unbalanced lines, and these are only suitable for communications over relatively short distances.

1.6.3 Transmission lines

At low data rates a piece of wire can be considered as a simple entity which conducts current and perhaps attenuates the signal resistively to some extent, and in which all components of the signal travel at the same speed as each other, but at higher rates it is necessary to consider the interconnect as a 'transmission line' in which reflections may be set up and where such factors as the characteristic impedance and terminating impedance of the line become important.

When considering a simple electrical circuit it is normal to assume that changes in voltage and current occur at the same time throughout the circuit, since the speed at which electricity travels down a wire is fast enough for this to be a reasonable assumption in many cases. When very long cables are involved, the time taken for an electrical signal to travel from one end to the other may approach a significant proportion of one cycle of the signal's waveform, and when the frequency of the electrical signal is high this is yet more likely since the cycle is short. Another way of considering this is to think in terms of the effective wavelength of the signal in its electrical form, which will be very long since the speed at which electricity travels in wire approaches the speed of light. When the wavelength of the signal in the wire becomes of the same order of magnitude as the length of the wire, transmission line issues may arise.

In a transmission line the ends of the line may be considered to be impedance discontinuities, that is points at which the impedance of the line changes from the line's *characteristic impedance* to the impedance of the *termination*. The characteristic impedance of a line is a difficult concept to grasp but it may be modelled as shown in Figure 1.13, being a combination of inductance and capacitance (and probably, in reality, also some resistance) which depends on the spacing between the conductors, their size and the type of insulation used. Characteristic

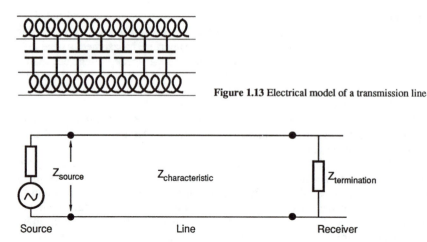

Figure 1.13 Electrical model of a transmission line

Matched transmission line when $Z_{source} = Z_{characteristic} = Z_{termination}$

Figure 1.14 Electrical characteristics of a matched transmission line

impedance has been defined as the input impedance of a line of infinite length[6]. The situation at the ends of a transmission line may be likened to what happens when a sound wave hits a wall – a portion of the power is reflected and a portion is absorbed. The wall represents an impedance discontinuity, and at such points sound energy is reflected. At certain frequencies standing wave modes will be set up in the room, at frequencies where multiples of half a wavelength equal one of the dimensions of the room, whereby the reflected wave combines constructively with the incident wave to produce points of maximum and minimum sound pressure within the room.

If an electrical transmission line is incorrectly terminated, reflections will be set up at the ends of the line, resulting in a secondary electrical wave travelling back down the line. This reflected wave may interfere with the transmitted wave, and in the case of a data signal may corrupt it if the reflected energy is high. The mismatch also results in a loss of power transferred to the next stage. If the line is correctly terminated the end of the line will not appear to be a discontinuity, the optimum power transfer will result at this point, and no reflections will be set up. A line is correctly terminated by ensuring that source and receiver impedances are the same as the characteristic impedance of the line (see Figure 1.14).

The upshot of all this for the purposes of this book is that long interconnects which carry high frequency audio or video data signals, often having bandwidths of many megahertz, may be subject to transmission line phenomena, and thus lines should normally be correctly terminated. The penalty for not doing so may be reduced signal strengths and erroneous data reception after some distance, and such situations can sometimes arise, especially when audio data signals are sent down existing cable runs which may pass through jackfields and different types of wire, each of which presents a change in characteristic impedance. In video environments people tend to be used to the concepts of transmission lines and correct termination, since video signals have always been subject to transmission line phenomena. In audio environments it may be a new concept, since analog audio signals do not contain high enough frequencies for such matters to become a problem, yet digital audio signals may. It is not recommended, for example, to parallel a number of receivers across one source line when dealing with digital signals, since the line will not then be correctly terminated and the signal level may also be considerably attenuated.

1.6.4 Cables

The types of cables used in serial audio and video interfaces vary from balanced screened, twisted pair (e.g. AES/EBU) to unbalanced coaxial links (e.g. Sony SDIF-2) and the characteristic impedances of these cables are different. Typical twisted pair audio cable, such as is used in many analog installations (and also put to use in digital links), tends to have a characteristic impedance of around 90–100 ohms, although this is not carefully controlled in manufacture since it normally does not matter for analog audio signals, whilst the more expensive 'star-quad' audio cable has a lower characteristic impedance of around 35 ohms. Typical coaxial cables used in video work have a characteristic impedance of 75 ohms, and other RF coaxial links use 50 ohm cable.

One problem with the twisted pair balanced line is that its electromagnetic radiation is poor compared with that of the coaxial link. A further advantage of the coaxial link is that its attenuation does not become severe until much higher frequencies than that of the twisted pair, and the speed of propagation along a coaxial link is significantly faster than along a twisted pair, resulting in smaller signal delays. But the twisted pair link is balanced whereas the coaxial link is not, and this makes it more immune to external interference, which is a great advantage.

Cable losses at high frequencies will clearly reduce the bandwidth of the interconnect, and the effect of this on the data signal is to slow the rise and fall times of the data edges, and to delay the point of transition between one state and the other by an amount which depends on the state of the data in the previous bit cell. The practical result of this is data link timing jitter, as proposed by Dunn[7], which may affect signal quality in D/A conversion if the clock is recovered from a portion of the data frame which is subject to a large degree of jitter. Links which suffer HF loss can often be equalized at the receiver to prop up the high frequency end of the spectrum, and this can help to accommodate longer interconnects which would otherwise fail. This matter is examined further in section 6.4.3.

Thus a number of factors combine to suggest that the type of cabling used in a digital interconnect is an important matter. A cable is required whose characteristic impedance matches that of the driver and receiver as closely as possible and is consistent along the length of the cable. The cable should have low loss at high frequencies, and the precise specification requires a knowledge of the bandwidth of the signal to be transferred. Cable manufacturers can usually quote such figures in specification sheets or on request, and it pays to study such matters, especially when recabling a large installation. More detailed discussion of these matters is contained within the sections of this book covering individual interfaces.

1.6.5 Connectors

The connectors used in digital interfacing fall into a number of distinct categories (see Figure 1.15). Firstly there are the unbalanced coaxial connectors, normally BNC type, which differ slightly depending on the characteristic impedance of the line; 50 ohm BNC connectors have a slightly larger central pin than 75 ohm

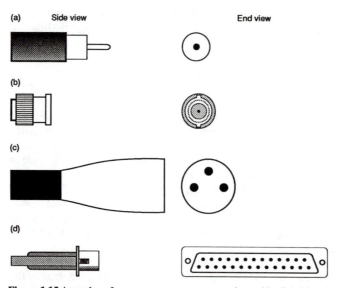

Figure 1.15 A number of connector types are commonly used in digital interfacing. (a) RCA phono connector. (b) BNC connector. (c) XLR connector. (d) D-type connector

connectors and can damage 75 ohm sockets if used inadvertently. RCA phono connectors, such as are often found in consumer hi-fi systems, are also used for unbalanced consumer digital audio interfaces using coaxial cable, although they are not proper coaxial connectors and do not have a controlled characteristic impedance. Both these connector types carry the data signal on the central pin and the shield on the outer ring.

The XLR-3 connector is used for one balanced digital audio interface, and it has its roots as an analog professional audio connector. The convention is for pin 1 to be the shield, pin 2 to be 'Data+', and pin 3 to be 'Data−'.

The D-sub type of connector stems from computer systems and remote control applications, and it has a number of individual pins arranged in two or three rows. The 9 pin D connector is often used for RS-422 communications, since it allows for two balanced sends and returns plus a ground, whereas some custom digital interfaces (either parallel or multichannel serial) use 25, 36 or even 50 pin D-type connectors.

These are not the only connectors used in audio and video interfacing, but they are the most common. Miscellaneous manufacturer-specific formats use none of these, Yamaha preferring the 8 pin DIN connector, for example, as its digital 'cascade' connector.

1.7 Optical fibres

Optical fibres are now playing a larger part in everyday data communications, and the cost:performance ratio of such fibre interconnects makes them a reasonable proposition when a large amount of data is to be carried over long distances. The key features of optical fibres are a large bandwidth, very low losses and immunity to interference, coupled with small size and flexibility. Also, when fibres are used as interconnects, devices are electrically isolated thus avoiding problems such as ground loops, shorts and crosstalk. Data is transferred over fibres by modulating a light source with the data, the light being carried within the fibre to an optical detector at the receiving end. The light source may be an LED or a laser diode, with the LED capable of operation up to a maximum of a few hundred megahertz at low power, whilst the laser is preferable in applications at higher frequencies or over longer distances.

1.7.1 Fibre principles

The typical construction of an optical fibre is shown in Figure 1.16, and light travels in the fibre as in a typical 'waveguide', by reflection at the boundaries. Total internal reflection occurs at the boundaries between the fibre core and its cladding, provided that the angle at which the light wave hits the boundary is shallower than a certain value called the *critical angle*, and thus the method of coupling the light source to the fibre is important in ensuring that light is propagated in the right way. Unless the fibre is exceptionally narrow, with a diameter approaching the wavelength of the light travelling along it (say between 4 and 10 μm), the light will travel along a number of paths – known as 'multimodes' – and thus the time taken for a source pulse to arrive at the receiver may vary slightly depending on the length of each path. This results in smearing of pulses in the time domain, and is known as *modal dispersion*, or, looked at in the frequency domain, it represents a reduction in the bandwidth of the link with increasing length. Up to distances of around 1.5 km the bandwidth decreases roughly linearly with distance (quoted in MHz per km), and after this in proportion to the square root of the length. Optical signals will also be attenuated with distance due to scattering by metal ions within the fibre, and by absorption due to water present within the structure. Losses are usually quoted in dB per km at a specified wavelength of light, and can be as low as 1 dB/km with high quality silica, graded index multimode fibres, or higher than 100 dB/km with plastic or ordinary glass cores. Clearly the high loss cables would be cheaper, and perhaps adequate for consumer applications in which the distances to be covered might be quite small.

Single mode fibres with very fine cores achieve very wide bandwidths with very low losses, and thus are suitable for use over long distances. Attenuations of around 0.5 dB/km are not uncommon with such fibres, which have only recently become feasible due to the development of suitable sources and connectors.

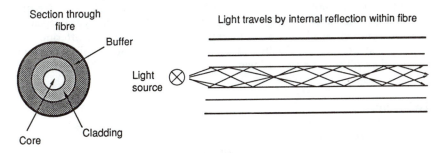

Figure 1.16 Cross section through a typical optical fibre, and mode of transmission

1.7.2 Light sources and connectors

LED (Light-Emitting Diode) light sources are made of gallium arsenide (GaAs) and can be doped to produce light with a wavelength between 800 and 1300 nm. The bandwidth of the radiated light is fairly wide, having a range of wavelengths of around 40 nm, which is another factor leading to greater losses over distance as the light of different wavelengths propagates over different modal paths within the fibre. The light from an LED source is incoherent (i.e. the phase and plane of the wavelets is random), and the angle over which it is radiated is quite wide. Since light radiated into the fibre at angles greater than a certain 'acceptance angle' will not be internally reflected it will effectively be lost in the cladding; thus the effectiveness of the coupling of light from an LED into the fibre is not good and only a few hundred microwatts of power can be transmitted.

An ILD (Injection Laser Diode) on the other hand produces coherent light of a similar wavelength to the LED, but over a narrower angle and with a narrower bandwidth (between around 1 and 3 nm in wavelength), thus providing better coupling of the light power to the fibre and resulting in less dispersion. Because of this, ILD drivers can be used for links which work in the gigahertz region whilst maintaining low losses.

Optical detectors are forms of photodiode which are very efficient at converting received light power into electrical current. Rise times of around 10 ns or less are achievable. The main problem with detectors is the distortion and noise they introduce, and different types of photodiode differ enormously in their S/N ratios. The so-called *avalanche photodiode* has a noise floor considerably lower than its counterpart, the *PIN diode*, and thus is preferable in critical applications. The *integrated detector/preamplifier* (IPD) provides amplification and detection of the light source in an integrated device, providing a higher output level than the other two and an improved S/N ratio.

The important features of connectors are their insertion loss and their return loss, these being respectively the amount of power lost due to the insertion of a connector into an otherwise unbroken link, and the amount of power reflected back in the direction of the source due to the presence of the connector. Typically insertion loss should be low (less than 1 dB), and return loss should be high (greater than 40 dB), for a reliable installation.

There are seven principal types of fibre optic connector, and it is not intended to cover each of them in detail here. For a comprehensive survey the reader is referred to Ajemian[8]. The so-called FDDI (Fibre Distributed Digital Interface) MIC connector is gaining widespread use in optical networks since it is a duplex connector (allows separate communication in two directions) with a typical insertion loss of 0.6 dB, whilst the SC connector is a popular 'snap-on' device developed by the Japanese NTT Corporation, available in both simplex and duplex forms, offering low insertion loss of around 0.25 dB and small size. The ST series of connectors, developed by AT&T, is also used widely.

References

1   RUMSEY, F.J. (1991) *Digital Audio Operations*, Focal Press, Oxford
2   WATKINSON, J.R. (1989) *The Art of Digital Audio*, Focal Press, Oxford
3   WATKINSON, J.R. (1990) *The Art of Digital Video*, Focal Press, Oxford

4   CONNOR, F.R. (1982) *Modulation,* Edward Arnold, London
5   EIA. *Industrial electronics bulletin no. 12.* EIA standard RS-422A. Electronics Industries Association, Engineering Dept., Washington DC
6   SINNEMA, W. (1979) *Electronic Transmission Technology*, Prentice-Hall, Englewood Cliffs, NJ
7   DUNN, J. (1992) Jitter: specification and assessment in digital audio equipment. Presented at the *93rd AES Convention, San Francisco,* 1–4 October. Audio Engineering Society
8   AJEMIAN, R.G. (1992) Fiber-optic connector considerations for professional audio. *Journal of the Audio Engineering Society*, vol. 40, no. 6, June, pp. 524–531

Bibliography

AJEMIAN, G. and GRUNDY, A. (1990) Fibre optics. The new medium for audio – a tutorial. *Journal of the Audio Engineering Society*, vol. 38, no. 3, March, pp. 160–175
TUGAL, D.A. and TUGAL, O. (1989) *Data Transmission*, 2nd edn, McGraw-Hill

Chapter 2

An introduction to digital audio and video

2.1 What is an audio signal?

Actual sounds are converted to electrical signals for convenience of handling, recording and conveying from one place to another. This is the job of the microphone. There are two basic types of microphone: those which measure the variations in air pressure due to sound, and those which measure the air velocity due to sound, although there are numerous practical types which are a combination of both.

The sound pressure or velocity varies with time and so does the output voltage of the microphone, in proportion. The output voltage of the microphone is thus an analogue of the sound pressure or velocity.

As sound causes no overall air movement, the average velocity of all sounds is zero, which corresponds to silence. As a result the bi-directional air movement gives rise to bipolar signals from the microphone, where silence is in the centre of the voltage range, and instantaneously negative or positive voltages are possible. Clearly the average voltage of all audio signals is also zero, and so when level is measured, it is necessary to take the modulus of the voltage, which is the job of the rectifier in the level meter. When this is done, the greater the amplitude of the audio signal, the greater the modulus becomes, and so a higher level is displayed.

Whilst the nature of an audio signal is very simple, there are many applications of audio, each requiring different bandwidth and dynamic range.

2.2 What is a video signal?

The goal of television is to allow a moving picture to be seen at a remote place. The picture is a two-dimensional image, which changes as a function of time. This is a three-dimensional information source where the dimensions are distance across the screen, distance down the screen and progression in time.

Whilst telescopes convey these three dimensions directly, this cannot be done with electrical signals or radio transmissions, which are restricted to a single parameter varying with time.

The solution in film and television is to convert the three-dimensional moving image into a series of still pictures, taken at the frame rate, and then, in television only, the two-dimensional images are scanned as a series of lines to produce a single voltage varying with time which can be recorded or transmitted.

2.3 Types of video

Figure 2.1 shows some of the basic types of analog colour video. Each of these types can, of course, exist in a variety of line standards. Since practical colour cameras generally have three separate sensors, one for each primary colour, an RGB system will exist at some stage in the internal workings of the camera, even if it does not emerge in that form. RGB consists of three parallel signals each having the same spectrum, and is used where the highest accuracy is needed, often for production of still pictures. Examples of this are paint systems and in computer-aided design (CAD) displays. RGB is seldom used for real time video recording; there is no standard RGB recording format for post-production or broadcast, although the IBA did build an experimental analog RGB recorder. As the red, green and blue signals directly represent part of the image, this approach is known as component video.

Some saving of bandwidth can be obtained by using colour difference working. The human eye relies on brightness to convey detail, and much less resolution is needed in the colour information. R, G and B are matrixed together to form a luminance (and monochrome compatible) signal Y, which has full bandwidth. The

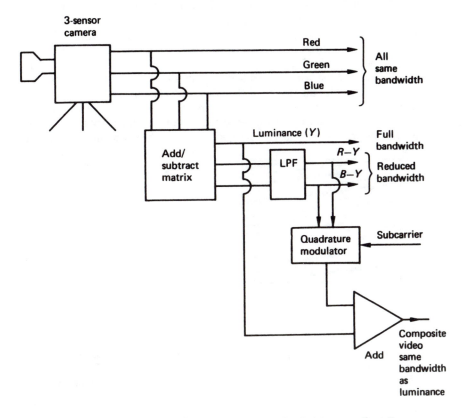

Figure 2.1 The major types of analog video. Red, green and blue signals emerge from the camera sensors, needing full bandwidth. If a luminance signal is obtained by a weighted sum of R, G and B, it will need full bandwidth, but the colour difference signals $R–Y$ and $B–Y$ need less bandwidth. Combining $R–Y$ and $B–Y$ in a subcarrier modulation scheme allows colour transmission in the same bandwidth as monochrome

matrix also produces two colour difference signals, $R–Y$ and $B–Y$, but these do not need the same bandwidth as Y; one-half or one-quarter will do depending on the application. Analog colour difference recorders such as Betacam and M II record these signals separately. The D-1 and D-5 formats record 525/60 or 625/50 colour difference signals after they have been digitized. Digital Betacam accepts the same signals but uses data reduction or compression. In casual parlance, colour difference formats are often called component formats to distinguish them from composite formats.

For colour television broadcast in a single channel, the PAL and NTSC systems interleave into the spectrum of a monochrome signal a subcarrier which carries two colour difference signals of restricted bandwidth. The subcarrier is intended to be invisible on the screen of a monochrome television set. A subcarrier-based colour system is generally referred to as composite video, and the modulated subcarrier is called chroma.

PAL uses the scanning standard of 625/50, whereas NTSC scans at 525/59.94, although the two colour systems have a few things in common. Analog composite recorders include the B and C formats, and the D-2 and D-3 formats which record PAL or NTSC digitally. A composite digital machine having analog inputs and outputs can replace a C-Format machine directly.

The terrestrial broadcast standards PAL and NTSC use 2:1 interlace. Figure 2.2(a) shows that in such a system, there is an odd number of lines in a frame, and the frame is split into two fields. The first field begins with a whole line and ends with a half line, and the second field begins with a half line, which allows it to interleave spatially with the first field. The field rate is intended to determine the flicker frequency, whereas the frame rate determines the bandwidth needed, which is thus halved along with the information rate. Information theory tells us that halving the information rate must reduce quality, and so the saving in bandwidth is

First field begins on a full   Second field begins on a half
line and ends on a half line   line and ends on a full line

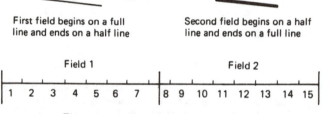

Field 1   Field 2

1   2   3   4   5   6   7 | 8   9   10 11 12 13 14 15

There must be an odd number of lines in a frame

Figure 2.2(a) 2:1 interlace

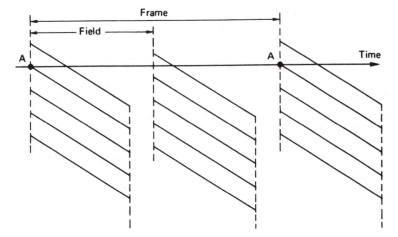

Figure 2.2(b) In an interlaced system, a given point A is only refreshed at frame rate, causing twitter on fine vertical detail

accompanied by a variety of effects. Figure 2.2(b) shows the spatial/temporal sampling points in a 2:1 interlaced system. If an object has a sharp horizontal edge, it will be present in one field but not in the next. The refresh rate of the edge will be reduced to frame rate, 25 Hz or 30 Hz, and becomes visible as twitter. Whilst the vertical resolution of a test card is maintained with interlace, apart from the twitter noted, the ability of an interlaced standard to convey motion is halved. In the light of what is now known, interlace causes degradation roughly proportional to bandwidth reduction, and so should not be considered for any future standards. Video interfaces have to accept the shortcomings inherent in the signals they convey. Their job is simply to pass on the signals without any further degradation.

2.4 What is a digital signal?

It is a characteristic of analog systems that degradations cannot be separated from the original signal, so nothing can be done about them. At the end of a system a signal carries the sum of all degradations introduced at each stage through which it passed. This sets a limit to the number of stages through which a signal can be passed before it is useless. Alternatively, if many stages are envisaged, each piece of equipment must be far better than necessary so that the signal is still acceptable at the end. The equipment will naturally be more expensive.

One of the vital concepts to grasp is that digital audio and video are simply alternative means of carrying the same information. An ideal digital recorder has the same characteristics as an ideal analog recorder: both of them are totally transparent and reproduce the original applied waveform without error. One need only compare high quality analog and digital equipment side by side with the same signals to realize how transparent modern equipment can be. Needless to say, in the real world, ideal conditions seldom prevail, so analog and digital equipment both fall short of the ideal. Digital equipment simply falls short of the ideal to a smaller extent than does analog and at lower cost, or, if the designer chooses, can have the same performance as analog at much lower cost.

Although there are a number of ways in which audio and video waveforms can be represented digitally, there is one system, known as pulse code modulation (PCM) which is in virtually universal use. Figure 2.3 shows how PCM works. Instead of being continuous, the time axis is represented in a discrete, or stepwise, manner. The waveform is not carried by continuous representation, but by measurement at regular intervals. This process is called sampling and the frequency with which samples are taken is called the sampling rate or sampling frequency F_s. The sampling rate is generally fixed and is not necessarily a function of any frequency in the signal, although in video it may be for convenience. If every effort is made to rid the sampling clock of jitter, or time instability, every sample will be made at an exactly even time step. Clearly if there is any subsequent timebase error, the instants at which samples arrive will be changed and the effect can be detected. If samples arrive at some destination with an irregular timebase, the effect can be eliminated by storing the samples temporarily in a memory and reading them out using a stable, locally generated clock. This process is called timebase correction and all properly engineered digital systems must use it. Clearly timebase error is not simply reduced; it can be totally eliminated. As a result there is little point measuring the wow and flutter or timebase error of a digital recorder; it doesn't have any. What happens is that the crystal clock in the timebase corrector measures the stability of the measuring set. It should be stressed that sampling is an analog process. Each sample still varies infinitely as the original waveform did.

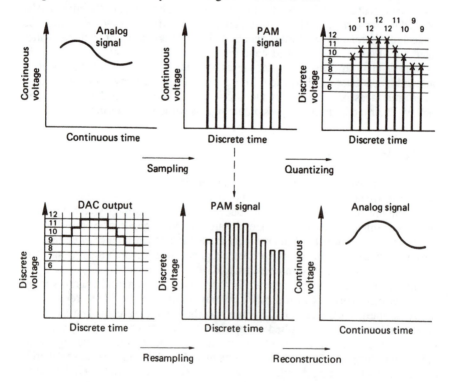

Figure 2.3 The major processes in PCM conversion. A/D conversion (top) and D/A conversion (bottom). Note that the quantizing step can be omitted to examine sampling and reconstruction independently of quantizing (dotted arrow)

Those who are not familiar with digital processes often worry that sampling takes away something from a signal because it is not taking notice of what happened between the samples. This would be true in a system having infinite bandwidth, but no analog signal can have infinite bandwidth. All analog signal sources from microphones, tape decks, cameras and so on have a frequency response limit, as indeed do our ears and eyes. When a signal has finite bandwidth, the rate at which it can change is limited, and the way in which it changes becomes predictable. When a waveform can only change between samples in one way, it is then only necessary to carry the samples and the original waveform can be reconstructed from them.

Figure 2.3 also shows that each sample is also discrete, or represented in a stepwise manner. The length of the sample, which will be proportional to the voltage of the waveform, is represented by a whole number. This process is known as quantizing and results in an approximation, but the size of the error can be controlled until it is negligible. If, for example, we were to measure the height of humans to the nearest metre, virtually all adults would register two metres high and obvious difficulties would result. These are generally overcome by measuring height to the nearest centimetre. Clearly there is no advantage in going further and expressing our height in a whole number of millimetres or even micrometres. The point is that an appropriate resolution can be found just as readily for audio or video, and greater accuracy is not beneficial. The link between quality and sample resolution is explored later in this chapter. The advantage of using whole numbers is that they are not prone to drift. If a whole number can be carried from one place to another without numerical error, it has not changed at all. By describing waveforms numerically, the original information has been expressed in a way which is better able to resist unwanted changes.

Essentially, digital systems carry the original waveform numerically. The number of the sample is an analogue of time, and the magnitude of the sample is an analogue of the signal voltage. As both axes of the waveform are discrete, the waveform can be accurately restored from numbers as if it were being drawn on graph paper. If we require greater accuracy, we simply choose paper with smaller squares. Clearly more numbers are required and each one could change over a larger range.

In simple terms, the waveform is conveyed in a digital recorder as if the voltage had been measured at regular intervals with a digital meter and the readings had been written down on a roll of paper. The rate at which the measurements were taken and the accuracy of the meter are the only factors which determine the quality, because once a parameter is expressed as a discrete number, a series of such numbers can be conveyed unchanged. Clearly in this example the handwriting used and the grade of paper have no effect on the information. The quality is determined only by the accuracy of conversion and is independent of the quality of the signal path.

In practical systems, binary numbers are used, as was explained in Chapter 1 which also showed that there are two ways in which binary signals can be used to carry samples. When each digit of the binary number is carried on a separate wire this is called parallel transmission. The state of the wires changes at the sampling rate. This approach is used in the parallel video interfaces, as video needs a relatively short word length: eight or ten bits. Using multiple wires is cumbersome where a long word length is in use, and a single wire can be used where successive digits from each sample are sent serially. This is the definition of pulse code modulation. Clearly the clock frequency must now be higher than the sampling rate. Whilst the transmission of audio by such a scheme is advantageous in that noise and timebase error have been eliminated, there is a penalty that a single high quality audio channel requires around one million bits per second. Digital audio could only come into use when such a data rate could be handled economically.

As a digital video channel requires of the order of two hundred million bits per second it is not surprising that digital audio equipment became common somewhat before digital video.

2.5 Why digital?

There are two main answers to this question, and it is not possible to say which is the most important, as it will depend on one's standpoint:

(a) The quality of reproduction of a well engineered digital system is independent of the medium and, in the absence of a compression scheme, depends only on the quality of the conversion processes.

(b) The conversion to the digital domain allows tremendous opportunities which were denied to analog signals.

Someone who is only interested in quality will judge the former the most relevant. If good quality convertors can be obtained, all of the shortcomings of analog recording and transmission can be eliminated to great advantage. One's greatest effort is expended in the design of convertors, whereas those parts of the system which handle data need only be workmanlike. Wow, flutter, timebase error, vector jitter, crosstalk, particulate noise, print-through, dropouts, modulation noise, HF squashing, azimuth error and interchannel phase errors are all history. When a digital recording is copied, the same numbers appear on the copy: it is not a dub, it is a clone. If the copy is indistinguishable from the original, there has been no generation loss. Digital recordings can be copied indefinitely without loss of quality through a digital interface.

In the real world everything has a cost, and one of the greatest strengths of digital technology is low cost. If copying causes no quality loss, recorders do not need to be far better than necessary in order to withstand generation loss. They need only be of adequate quality on the first generation if that quality is then maintained. There is no need for the great size and extravagant tape consumption of professional analog recorders. When the information to be recorded is discrete numbers, they can be packed densely on the medium without quality loss. Should some bits be in error because of noise or dropout, error correction can restore the original value. Digital recordings take up less space than analog recordings for the same or better quality. Tape costs are far less and storage costs are reduced.

Digital circuitry costs less to manufacture. Switching circuitry which handles binary can be integrated more densely than analog circuitry. More functionality can be put in the same chip. Analog circuits are built from a host of different component types which have a variety of shapes and sizes and are costly to assemble and adjust. Digital circuitry uses standardized component outlines and is easier to assemble on automated equipment. Little if any adjustment is needed. Once audio or video signals are in the digital domain, they become data, and apart from the need to be reproduced at the correct sampling rate, are indistinguishable from any other type of data. Systems and techniques developed in other industries for other purposes can be used for audio and video. Computer equipment is available at low cost because the volume of production is far greater than that of professional audio equipment. Disk drives and memories developed for computers can be put to use in audio and video products. A word processor adapted to handle audio samples becomes a workstation. If video data is handled the result is a non-linear editor. There seems to be little point in waiting for a tape to wind when a disk head can access data in milliseconds. The concept of editing by cutting tape causes hysterics in anyone in the computer industry. Imagine editing a text file in a word processor by taking scissors to a floppy disk! The difficulty of locating the edit point and the irrevocable nature of tape-cut editing are hardly worth considering when the edit point can be located by viewing the material on a screen or by listening at any speed to audio from a memory. The edit can be simulated and trimmed before it is made permanent.

Communications networks developed to handle data can happily carry digital audio and video over indefinite distances without quality loss. Digital audio broadcasting (DAB) makes use of these techniques to eliminate the interference, fading and multipath reception problems of analog broadcasting. At the same time, more efficient use is made of available bandwidth. Digital broadcasting techniques are now being applied to television signals, with DVB being used in Europe and possibly other parts of the world too.

Digital equipment can have self-diagnosis programs built-in. The machine points out its own failures. The days of chasing a signal with an oscilloscope are over. Even if a faulty component in a digital circuit could be located with such a primitive tool, it is virtually impossible to replace a chip having 60 pins soldered through a six-layer circuit board. The cost of finding the fault may be more than the board is worth. Routine, mind-numbing adjustment of analog circuits to counteract drift is no longer needed. The cost of maintenance falls. A small operation may not need maintenance staff at all; a service contract is sufficient. A larger organization will still need maintenance staff, but they will be fewer in number and more highly skilled.

As a result of the above, the cost of ownership of digital equipment is less than that of analog. Debates about quality are academic except where compression is concerned; analog equipment can no longer compete economically, and it will dwindle away as surely as the transistor once replaced the vacuum-tube in electronics and the turbine replaced the piston engine in aviation.

2.6 The information content of an analog signal

Any analog signal source can be characterized by a given useful bandwidth and signal-to-noise ratio. Video signals have very wide bandwidth extending over several MHz and require only 50 dB or so S/N ratio whereas audio signals require only 20 kHz but need much better S/N ratio. If a well-engineered digital channel having a wider bandwidth and a greater signal-to-noise ratio is put in series with such a source, it is only necessary to set the levels correctly and the signal is then subject to no loss of information whatsoever.

Provided the digital clipping level is above the largest input signal, the digital noise floor is below the inherent noise in the input signal and the low- and high-frequency response of the digital channel extends beyond the frequencies in the input signal, then the digital channel is a 'wider window' than the input signal needs and its extremities cannot be explored by that signal. As a result there is no test known which can reliably tell whether or not the digital system was or was not present, unless, of course, it is deficient in some quantifiable way.

In audio the wider-window effect is obvious on certain Compact Discs which are made from analog master tapes. The CD player faithfully reproduces the tape hiss, dropouts and HF squashing of the analog master, which render the entire CD mastering and reproduction system transparent by comparison.

On the other hand, if an analog source can be found which has a wider window than the digital system, then the presence of the digital system will be evident either due to the reduction in bandwidth or the reduction in dynamic range.

2.7 Introduction to conversion

There are a number of ways in which an audio waveform can be digitally represented, but the most useful and therefore common is PCM which was introduced in Chapter 1. The input is a continuous-time, continuous-voltage waveform, and this is converted into a discrete-time, discrete-voltage format by a combination of sampling and quantizing. As these two processes are orthogonal (a 64 dollar word meaning at right angles to one another) they are totally independent and can be performed in either order. Figure 2.4(a) shows an analog sampler preceding a quantizer, whereas Figure 2.4(b) shows an asynchronous quantizer preceding a digital sampler. Ideally, both will give the same results; in practice each suffers from different deficiencies. Both approaches will be found in real equipment. In video convertors operating speed is a priority and the flash convertor is universally used. In the flash convertor the quantizing step is performed first and the quantized signal is subsequently sampled. In audio the sample accuracy is a priority and sampling first has the effect of freezing the signal voltage to allow time for an accurate quantizing process.

The independence of sampling and quantizing allows each to be discussed quite separately in some detail, prior to combining the processes for a full understanding of conversion.

2.7.1 Sampling and aliasing

Sampling is no more than periodic measurement, and it will be shown here that there is no theoretical need for sampling to be audible. Practical equipment may, of course, be less than ideal, but, given good engineering practice, the ideal may be approached quite closely.

Sampling must be precisely regular, because the subsequent process of timebase correction assumes a regular original process. The sampling process originates with a pulse train which is shown in Figure 2.5(a) to be of constant amplitude and period. The signal waveform amplitude-modulates the pulse train in much the same way as the carrier is modulated in an AM radio transmitter. One must be careful to avoid over-modulating the pulse train as shown in Figure 2.5(b) and this is helped by applying a DC offset to the analog waveform so that, in audio, silence corresponds to a level half-way up the pulses as in Figure 2.5(c). This approach will also be used in colour difference signals where blanking level will be shifted up to the midpoint of the scale. Clipping due to any excessive input level will then be symmetrical.

In the same way that AM radio produces sidebands or images above and below

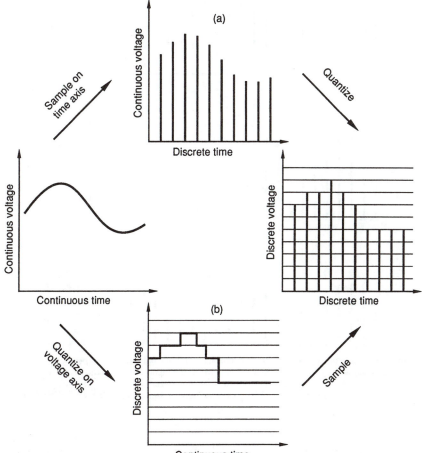

Figure 2.4 Since sampling and quantizing are orthogonal, the order in which they are performed is not important. At (a) sampling is performed first and the samples are quantized. This is common in audio convertors. At (b) the analog input is quantized into an asynchronous binary code. Sampling takes place when this code is latched on sampling clock edges. This approach is universal in video convertors

the carrier, sampling also produces sidebands; although the carrier is now a pulse train and has an infinite series of harmonics as can be seen in Figure 2.6(a). The sidebands in Figure 2.6(b) repeat above and below each harmonic of the sampling rate.

The sampled signal can be returned to the continuous-time domain simply by passing it into a low-pass filter. This filter has a frequency response which prevents the images from passing, and only the baseband signal emerges, completely unchanged.

If an input is supplied having an excessive bandwidth for the sampling rate in use, the sidebands will overlap, and the result is aliasing, where certain output frequencies are not the same as their input frequencies but become difference frequencies. It will be seen from Figure 2.6(c) that aliasing occurs when the input frequency

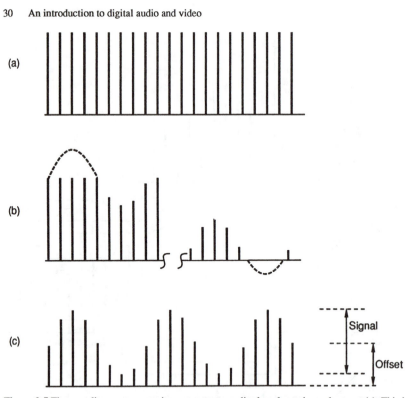

Figure 2.5 The sampling process requires a constant amplitude pulse train as shown at (a). This is amplitude modulated by the waveform to be sampled. If the input waveform has excessive amplitude or incorrect level, the pulse train clips as shown at (b). For an audio waveform, the greatest signal level is possible when an offset of half the pulse amplitude is used to centre the waveform as shown at (c)

exceeds half the sampling rate, and this derives the most fundamental rule of sampling, first stated by Shannon in the West and at about the same time by Kotelnikov in Russia. This states that the sampling rate must be at least twice the highest input frequency.

In addition to the low-pass filter needed at the output to return to the continuous-time domain, a further low-pass filter is needed at the input to prevent aliasing. If input frequencies of more than half the sampling rate cannot reach the sampler, aliasing cannot occur.

Whilst aliasing has been described above in the frequency domain, it can equally be described in the time domain. In Figure 2.7(a) the sampling rate is obviously adequate to describe the waveform, but in Figure 2.7(b) it is inadequate and aliasing has occurred.

Aliasing is commonly seen on television and in the cinema, owing to the relatively low frame rates used. With a frame rate of 24 Hz, a film camera will alias on any object changing at more than 12 Hz. Such objects include the spokes of stagecoach wheels, especially when being chased by Indians. When the spoke-passing frequency reaches 24 Hz the wheels appear to stop. Aliasing partly explains the inability of opinion polls to predict the results of elections. Aliasing does, however, have useful applications, including the stroboscope, which makes rotating

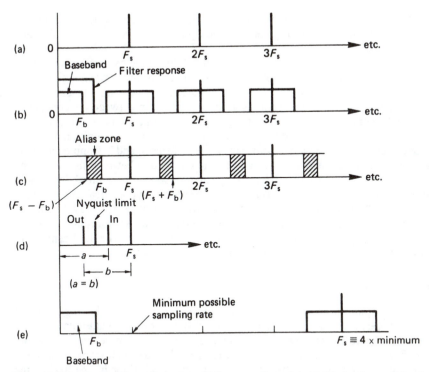

Figure 2.6 (a) Spectrum of sampling pulses. (b) Spectrum of samples. (c) Aliasing due to sideband overlap. (d) Beat-frequency production. (e) Four-times oversampling

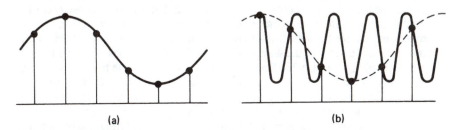

Figure 2.7 At (a) the sampling rate is adequate to reconstruct the original signal. At (b) the sampling rate is inadequate, and reconstruction produces the wrong waveform (dotted). Aliasing has taken place

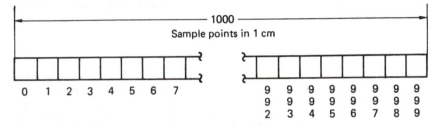

Figure 2.8 If the above spatial sampling arrangement of 1000 points per centimetre is scanned in 1 millisecond, the sampling rate will become 1 megahertz

machinery appear stationary, and the sampling oscilloscope, which can display periodic waveforms of much greater frequency than the sweep speed of the tube normally allows.

In television systems the input image which falls on the camera sensor will be continuous in time, and continuous in two spatial dimensions corresponding to the height and width of the sensor. All three of these continuous dimensions will be sampled in a digital system. There is a direct connection between the concept of temporal sampling, where the input signal changes with respect to time at some frequency and is sampled at some other frequency, and spatial sampling, where an image changes a given number of times per unit distance and is sampled at some other number of times per unit distance. The connection between the two is the process of scanning. Temporal frequency can be obtained by multiplying spatial frequency by the speed of the scan. Figure 2.8 shows a hypothetical image sensor which has 1000 discrete sensors across a width of 1 centimetre. The spatial sampling rate of this sensor is thus 1000 per centimetre. If the sensors are measured sequentially during a scan which takes 1 millisecond to go across the 1 centimetre width, the result will be a temporal sampling rate of 1 MHz.

2.7.2 Reconstruction

If ideal low-pass anti-aliasing and anti-image filters are assumed, having a vertical cut-off slope at half the sampling rate, an ideal spectrum shown at Figure 2.9(a) is obtained. Figure 2.9(b) shows that the impulse response of a phase linear ideal low-pass filter is a sin x/x waveform in the time domain. Such a waveform passes through zero volts periodically. If the cut-off frequency of the filter is one-half of the sampling rate, the impulse passes through zero *at the sites of all other samples*. Thus at the output of such a filter, the voltage at the centre of a sample is due to that sample alone, since the value of *all* other samples is zero at that instant. In other words the continuous time output waveform must join up the tops of the input samples. In between the sample instants, the output of the filter is the sum of the contributions from many impulses, and the waveform smoothly joins the tops of the samples. If the time domain is being considered, the anti-image filter of the frequency domain can equally well be called the reconstruction filter. It is a consequence of the band-limiting of the original anti-aliasing filter that the filtered analog waveform could only travel between the sample points in one way. As the reconstruction filter has the same frequency response, the reconstructed output waveform must be identical to the original band-limited waveform prior to sampling. It follows that sampling need not be audible. The reservations expressed by some journalists about 'hearing the gaps between the samples' clearly have no foundation whatsoever. A rigorous mathematical proof of the above has been available since the 1930s, when PCM was invented, and can also be found in Betts[1].

2.7.3 Filter design

The ideal filter with a vertical 'brick-wall' cut-off slope is difficult to implement. As the slope tends to vertical, the delay caused by the filter goes to infinity: the quality is marvellous but you do not live to measure it. In practice, a filter with a finite slope is accepted as shown in Figure 2.10, and the sampling rate has to be raised a little to prevent aliasing. There is no absolute factor by which the sampling rate must be raised; it depends upon the filters which are available.

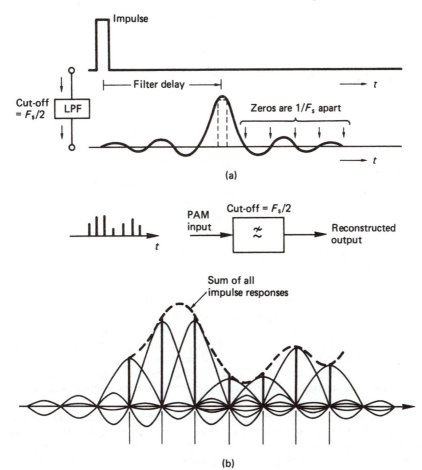

(a)

(b)

Figure 2.9 The impulse response of a low-pass filter which cuts off at $F_s/2$ has zeros at $1/F_s$ spacing which correspond to the position of adjacent samples, as shown at (b). The output will be a signal which has the value of each sample at the sample instant, but with smooth transitions from sample to sample

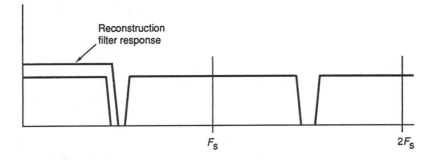

Figure 2.10 As filters with finite slope are needed in practical systems, the sampling rate is raised slightly beyond twice the highest frequency in the baseband

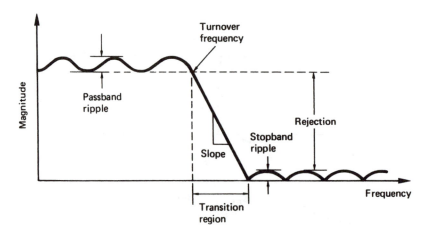

Figure 2.11 The important features and terminology of low-pass filters used for anti-aliasing and reconstruction

It is not easy to specify such filters, particularly the amount of stopband rejection needed. The amount of aliasing resulting would depend on, among other things, the amount of out-of-band energy in the input signal. This is seldom a problem in video, but can warrant attention in audio where overspecified bandwidths are sometimes found. As a further complication, an out-of-band signal will be attenuated by the response of the anti-aliasing filter to that frequency, but the residual signal will then alias, and the reconstruction filter will reject it according to its attenuation at the new frequency to which it has aliased.

It could be argued that the reconstruction filter is unnecessary in audio, since all of the images are outside the range of human hearing. However, the slightest non-linearity in subsequent stages would result in gross intermodulation distortion. The possibility of damage to tweeters and beating with the bias systems of analog tape recorders must also be considered. It would, however, be acceptable to bypass one of the filters involved in a copy from one digital machine to another via the analog domain, although a digital transfer is of course to be preferred. In video the filters are essential to constrain the bandwidth of the signal to the allowable broadcast channel width.

The nature of the filters used has a great bearing on the subjective quality of the system. Entire books have been written about analog filters, so they will only be treated briefly here.

Figure 2.11 shows the terminology used to describe the common elliptic low-pass filter. These filters are popular because they can be realized with fewer components than other filters of similar response. It is a characteristic of these elliptic filters that there are ripples in the passband and stopband. In much equipment the anti-aliasing filter and the reconstruction filter will have the same specification, so that the passband ripple is doubled with a corresponding increase in dispersion. Sometimes slightly different filters are used to reduce the effect.

It is difficult to produce an analog filter with low distortion. Passive filters using inductors suffer non-linearity at high levels due to the *B/H* curve of the cores. Active filters can simulate inductors which are linear using op-amp techniques, but they tend to suffer non-linearity at high frequencies where the falling open-loop gain

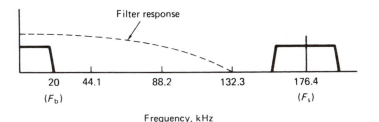

Frequency, kHz

Figure 2.12 In this 4-times oversampling system, the large separation between baseband and sidebands allows a gentle roll-off reconstruction filter to be used

reduces the effect of feedback. Active filters can also contribute noise, but this is not necessarily a bad thing in controlled amounts, since it can act as a dither source.

It is instructive to examine the phase response of such filters. Since a sharp cut-off is generally achieved by cascading many filter sections which cut at a similar frequency, the phase responses of these sections will accumulate. The phase may start to leave linearity at a fraction of the cut-off frequency and by the time this is reached, the phase may have completed several revolutions. Meyer[2] suggests that these phase errors are audible and that equalization is necessary. In video, phase linearity is essential as otherwise the different frequency components of a contrast step are smeared across the screen. An advantage of linear phase filters is that ringing is minimized, and there is less possibility of clipping on transients.

It is possible to construct a ripple-free phase-linear filter with the required stopband rejection[3,4], but the design effort and component complexity result in expense, and the filter might drift out of specification as components age. The effort may be more effectively directed towards avoiding the need for such a filter. Much effort can be saved in analog filter design by using oversampling. As shown in Figure 2.12 a high sampling rate produces a large spectral gap between the baseband and the first lower sideband. The anti-aliasing and reconstruction filters need only have a gentle roll-off, causing minimum disturbance to phase linearity in the baseband, and the Butterworth configuration, which does not have ripple or dispersion, can be used. The penalty of oversampling is that an excessive data rate results. It is necessary to reduce the rate using a digital low pass filter (LPF). Digital filters can be made perfectly phase linear and, using LSI, can be inexpensive to construct. The superiority of oversampling convertors means that they have become universal in audio and appear set to do so in video.

2.7.4 Sampling clock jitter

The instants at which samples are taken in an A/D convertor (ADC) and the instants at which D/A convertors (DACs) make conversions must be evenly spaced, otherwise unwanted signals can be added to the waveform. Figure 2.13(a) shows the effect of sampling clock jitter on a sloping waveform. Samples are taken at the wrong times. When these samples have passed through a system, the timebase correction stage prior to the DAC will remove the jitter, and the result is shown in Figure 2.13(b). The magnitude of the unwanted signal is proportional to the slope of the signal waveform and so increases with frequency. The nature of the unwanted signal depends on the spectrum of the jitter. If the jitter is random, the effect is noise-like and relatively benign unless the amplitude is excessive. Clock jitter is, however,

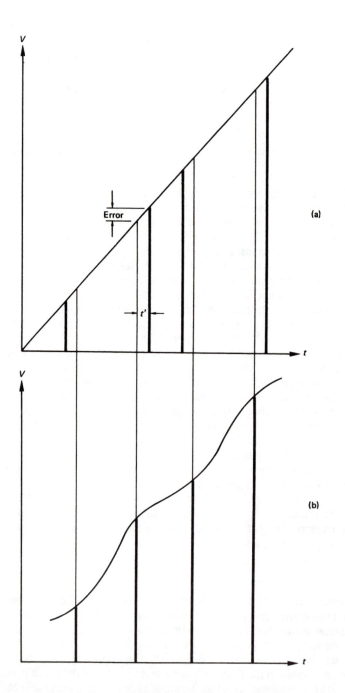

Figure 2.13 The effect of sampling timing jitter on noise. (a) A ramp sampled with jitter has an error proportional to the slope. (b) When the jitter is removed by later circuits, the error appears as noise added to samples. The superimposition of jitter may also be considered as a modulation process, as discussed in Chapter 6

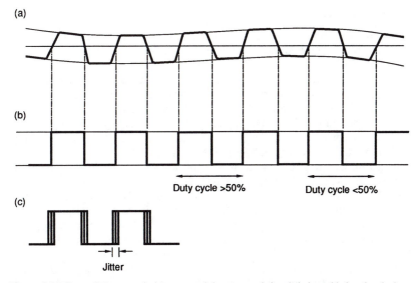

(a)

(b)

Duty cycle >50%   Duty cycle <50%

(c)

Jitter

Figure 2.14 Crosstalk in transmission can result in unwanted signals being added to the clock waveform. It can be seen here that a low frequency interference signal affects the slicing of the clock and causes a periodic jitter.

not necessarily random. Figure 2.14 shows that one source of clock jitter is crosstalk on the clock signal. The unwanted additional signal changes the time at which the sloping clock signal appears to cross the threshold voltage of the clock receiver. The threshold itself may be changed by ripple on the clock receiver power supply. There is no reason why these effects should be random; they may be periodic and potentially discernible.

The allowable jitter is measured in picoseconds in both audio and video signals, as shown in Figure 2.13, and clearly steps must be taken to eliminate it by design. Convertor clocks must be generated from clean power supplies which are well decoupled from the power used by the logic because a convertor clock must have a good signal-to-noise ratio. If an external clock is used, it cannot be used directly, but must be fed through a well-damped phase-locked loop which will filter out the jitter. The external clock signal is sometimes fed into the clean circuitry using an optical coupler to improve isolation.

Although it has been documented for many years, attention to control of clock jitter is not as great in actual audio hardware as it might be. It accounts for much of the slight audible differences between convertors reproducing the same data. A well-engineered convertor should substantially reject jitter on an external clock and should sound the same when reproducing the same data irrespective of the source of the data. A remote convertor which sounds different when reproducing, for example, the same Compact Disc via the digital outputs of a variety of CD players is simply not well engineered and should be rejected. Similarly if the effect of changing the type of digital cable feeding the convertor can be heard, the unit is a dud. Unfortunately many consumer external DACs fall into this category, as the steps outlined above have not been taken.

Jitter tends to be less noticeable on digital video signals and is generally not an issue until it becomes great enough to cause data errors.

2.7.5 Aperture effect

The reconstruction process of Figure 2.9 only operates exactly as shown if the impulses are of negligible duration. In many convertors this is not the case, and many keep the analog output constant until a different sample value is input and produce a waveform which is more like a staircase than a pulse train. In this case the pulses have effectively been extended in width to become equal to the sample period. This is known as a zero-order hold system and has a 100% aperture ratio. Note that the aperture effect is not apparent in a track-hold system; the holding period is for the convenience of the quantizer which outputs a value corresponding to the input voltage at the instant hold mode was entered.

Whereas pulses of negligible width have a uniform spectrum, which is flat within the audio band, pulses of 100% aperture have a sin x/x spectrum which is shown in Figure 2.15. The frequency response falls to a null at the sampling rate, and as a result is about 4 dB down at the edge of the baseband. If the pulse width is stable, the reduction of high frequencies is constant and predictable, and an appropriate equalization circuit can render the overall response flat once more. An alternative is to use resampling whose effect is shown in Figure 2.16. Resampling passes the zero-order hold waveform through a further synchronous sampling stage which consists of an analog switch which closes briefly in the centre of each sample period. The output of the switch will be pulses which are narrower than the original. If, for example, the aperture ratio is reduced to 50% of the sample period, the first response null is now at twice the sampling rate, and the loss at the edge of the audio band is reduced. As the figure shows, the frequency response becomes flatter as the aperture ratio falls. The process should not be carried too far, as with very small aperture ratios there is little energy in the pulses and noise can be a problem. A practical limit is around 12.5% where the frequency response is virtually ideal.

The aperture effect will show up in many aspects of television. Lenses have finite modulation transfer functions, such that a very small object becomes spread in the image. The image sensor will also have a finite aperture function. In tube cameras, the beam will have a finite radius, and will not necessarily have a uniform energy distribution across its diameter. In CCD cameras, the sensor is split into elements which may almost touch in some cases. The element integrates light falling on its surface, and so will have a rectangular aperture. In both cases there will be a roll-off of higher spatial frequencies.

It is highly desirable to prevent spatial aliasing, since the result is visually irritating. In tube cameras the aliasing will be in the vertical dimension only, since the horizontal dimension is continuously scanned. Such cameras seldom attempt to prevent vertical aliasing. CCD sensors can, however, alias in both horizontal and vertical dimensions, and so an anti-aliasing optical filter is generally fitted between the lens and the sensor. This takes the form of a plate which diffuses the image formed by the lens. Such a device can never have a sharp cut-off nor will the aperture be rectangular. The aperture of the anti-aliasing plate is in series with the aperture effect of the CCD elements, and the combination the two effectively prevents spatial aliasing, and generally gives a good balance between horizontal and vertical resolution, allowing the picture a natural appearance.

Conventional tube cameras generally have better horizontal resolution, and produce vertical aliasing which has a similar spatial frequency to real picture information, but which lacks realism. With a conventional approach, there are effectively two choices. If aliasing is permitted, the theoretical information rate of

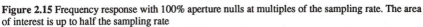

Figure 2.15 Frequency response with 100% aperture nulls at multiples of the sampling rate. The area of interest is up to half the sampling rate

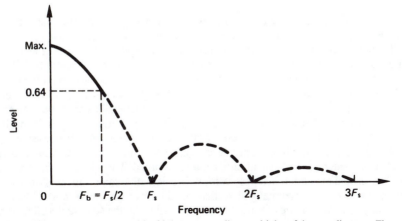

(a)

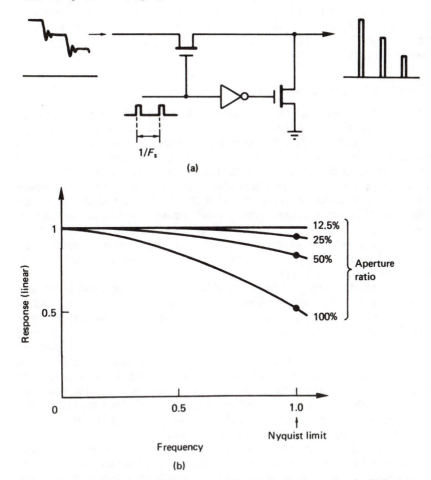

(b)

Figure 2.16 (a) A resampling circuit eliminates transients and reduces aperture ratio. (b) Response of various aperture ratios

the system can be approached. If aliasing is prevented, realizable anti-aliasing filters cannot sharp cut, and the information conveyed is below system capacity.

These considerations also apply at the television display. The display must filter out spatial frequencies above one-half the sampling rate. In a conventional CRT this means that an optical filter should be fitted in front of the screen to render the raster invisible. Again the aperture of a simply realizable filter would attenuate too much of the wanted spectrum, and so the technique is not used. The technique of spot wobble was most effective in reducing raster visibility in monochrome television sets without affecting horizontal resolution, and its neglect remains a mystery.

As noted, in conventional tube cameras and CRTs the horizontal dimension is continuous, whereas the vertical dimension is sampled. The aperture effect means that the vertical resolution in real systems will be less than sampling theory permits, and to obtain equal horizontal and vertical resolutions a greater number of lines is necessary. The magnitude of the increase is described by the so-called Kell factor[5], although the term factor is a misnomer since it can have a range of values depending on the apertures in use and the methods used to measure resolution[6]. In digital video, sampling takes place in horizontal and vertical dimensions, and the Kell parameter becomes unnecessary. The outputs of digital systems will, however, be displayed on raster scan CRTs, and the Kell parameter of the display will then be effectively in series with the other system constraints.

2.7.6 Choice of audio sampling rate

The Nyquist criterion is only the beginning of the process which must be followed to arrive at a suitable sampling rate. The slope of available filters will compel designers to raise the sampling rate above the theoretical Nyquist rate. For consumer products, the lower the sampling rate the better, since the cost of the medium is directly proportional to the sampling rate: thus sampling rates near to twice 20 kHz are to be expected. For professional products, there is a need to operate at variable speed for pitch correction. When the speed of a digital recorder is reduced, the offtape sampling rate falls, and Figure 2.17 shows that with a minimal sampling rate the first image frequency can become low enough to pass the reconstruction filter. If the sampling frequency is raised without changing the response of the filters, the speed can be reduced without this problem. It follows that variable-speed recorders must use a higher sampling rate.

In the early days of digital audio research, the necessary bandwidth of about 1 megabit per second per audio channel was difficult to store. Disk drives had the bandwidth but not the capacity for long recording time, so attention turned to video recorders. These were adapted to store audio samples by creating a pseudo-video waveform which could convey binary as black and white levels[7]. The sampling rate of such a system is constrained to relate simply to the field rate and field structure of the television standard used, so that an integer number of samples can be stored on each usable TV line in the field. Such a recording can be made on a monochrome recorder, and these recordings are made in two standards, 525 lines at 60 Hz and 625 lines at 50 Hz. Thus it is possible to find a frequency which is a common multiple of the two and also suitable for use as a sampling rate.

The allowable sampling rates in a pseudo-video system can be deduced by multiplying the field rate by the number of active lines in a field (blanked lines cannot be used) and again by the number of samples in a line. By careful choice of parameters it is possible to use either 525/60 or 625/50 video with a sampling rate of 44.1 kHz.

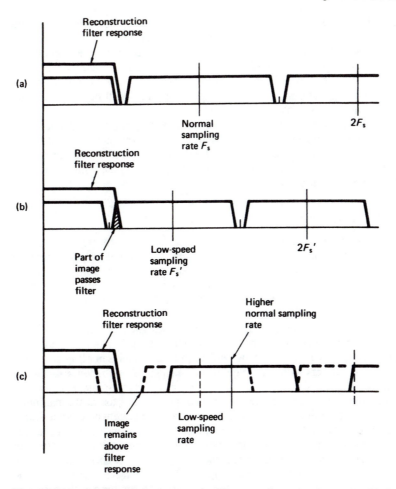

Figure 2.17 At normal speed, the reconstruction filter correctly prevents images entering the baseband, as at (a). (b) When speed is reduced, the sampling rate falls, and a fixed filter will allow part of the lower sideband of the sampling frequency to pass. (c) If the sampling rate of the machine is raised, but the filter characteristics remain the same, the problem can be avoided

In 60 Hz video, there are 35 blanked lines, leaving 490 lines per frame, or 245 lines per field for samples. If three samples are stored per line, the sampling rate becomes $60 \times 245 \times 3 = 44.1$ kHz

In 50 Hz video, there are 37 lines of blanking, leaving 588 active lines per frame, or 294 per field, so the same sampling rate is given by $50 \times 294 \times 3 = 44.1$ kHz.

The sampling rate of 44.1 kHz came to be that of the Compact Disc. Even though CD has no video circuitry, the equipment used to make CD masters is video based and determines the sampling rate.

For landlines to FM stereo broadcast transmitters having a 15 kHz audio bandwidth, the sampling rate of 32 kHz is more than adequate, and has been in use for some time in the United Kingdom and Japan. This frequency is also in use in the NICAM 728 stereo TV sound system. The professional sampling rate of 48 kHz was proposed as having a simple relationship to 32 kHz, being far enough above 40 kHz for variable-speed operation, and having a simple relationship with PAL video timing which would allow digital video recorders to store the convenient number of 960 audio samples per video field. This is the sampling rate used by all of the professional DVTR formats[8]. The field rate offset of NTSC does not easily relate to any of the above sampling rates, and requires special handling which will be discussed further in this book.

Although in a perfect world the adoption of a single sampling rate might have had virtues, for practical and economic reasons digital audio now has essentially three rates to support: 32 kHz for broadcast, 44.1 kHz for CD, and 48 kHz for professional use[9]. Variations of the RDAT format will support all of these rates, although the 32 kHz version is uncommon.

2.7.7 Choice of video sampling rate

Component or colour difference signals are used primarily for post-production work where quality and flexibility are paramount. In a digital colour difference system, the analog input video will be converted to the digital domain and will generally remain there whilst being operated on in digital effects units. The finished work will then be returned to the analog domain for transmission. The number of conversions is relatively few.

In contrast, composite digital video recorders are designed to work in an existing analog environment and so the number of conversions a signal passes through will be greater. It follows that composite will require a higher sampling rate than component so that filters with a more gentle slope can be used to reduce the build-up of response ripples.

In colour difference working, the important requirement is for image manipulation in the digital domain. This is facilitated by a sampling rate which is a multiple of line rate because then there is a whole number of samples in a line and samples are always in the same position along the line and can form neat columns. A practical difficulty is that the line period of the 525 and 625 systems is slightly different. The problem was overcome by the use of a sampling clock which is an integer multiple of both line rates. The rate of 13.5 MHz is sufficient for the requirements of sampling theory, yet allows a whole number of samples in both line standards with a common clock frequency.

The colour difference signals have only half the bandwidth of the luminance and so can be sampled at one-half the luminance rate, i.e. 6.75 MHz. Extended and high definition systems require proportionately higher rates.

In composite video, the most likely process to be undertaken in the digital domain is decoding to components. This is performed, for example, in video recorders running in slow motion. Separation of luminance and chroma in the digital domain is simpler if the sampling rate is a multiple of the subcarrier frequency. Whilst three times the frequency of subcarrier is adequate from a sampling theory standpoint, this does require relatively steep filters, so as a practical matter four times subcarrier is used.

2.7.8 Quantizing

Quantizing is the process of expressing some infinitely variable quantity by discrete or stepped values. Quantizing turns up in a remarkable number of everyday guises.

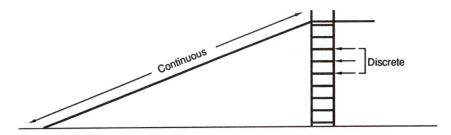

Figure 2.18 An analog parameter is continuous whereas a quantized parameter is restricted to certain values. Here the sloping side of a ramp can be used to obtain any height whereas a ladder only allows discrete heights

Figure 2.18 shows that an inclined ramp enables infinitely variable height to be achieved, whereas a step-ladder allows only discrete heights. A step-ladder quantizes height. When accountants round off sums of money to the nearest pound or dollar they are quantizing.

In audio the values to be quantized are infinitely variable voltages from an analog source. Strict quantizing is a process which is restricted to the signal amplitude domain only. For the purpose of studying the quantizing of a single sample, time is assumed to stand still. This is achieved in practice either by the use of a track-hold circuit or the adoption of a quantizer technology which operates before the sampling stage.

Figure 2.19(a) shows that the process of quantizing divides the signal voltage range into quantizing intervals Q. In applications such as telephony these may be of differing size, but for digital audio and video the quantizing intervals are made as identical as possible. If this is done, the binary numbers which result are truly proportional to the original analog voltage, and the digital equivalents of mixing and gain changing can be performed by adding and multiplying sample values. If the quantizing intervals are unequal this cannot be done. When all quantizing intervals are the same, the term uniform quantizing is used. The term linear quantizing will be found, but this is, like military intelligence, a contradiction in terms.

The term LSB (least significant bit) will also be found in place of quantizing interval in some treatments, but this is a poor term because quantizing is not always used to create binary values and because a bit can only have two values. In studying quantizing we wish to discuss values smaller than a quantizing interval, but a fraction of an LSB is a contradiction in terms.

Whatever the exact voltage of the input signal, the quantizer will determine the quantizing interval in which it lies. In what may be considered a separate step, the quantizing interval is then allocated a code value which is typically some form of binary number. The information sent is the number of the quantizing interval in which the input voltage lay. Exactly where that voltage lay within the interval is not conveyed, and this mechanism puts a limit on the accuracy of the quantizer. When the number of the quantizing interval is converted back to the analog domain, it will result in a voltage at the centre of the quantizing interval as this minimizes the magnitude of the error between input and output. The number range is limited by the word length of the binary numbers used. In a sixteen-bit system commonly used for audio, 65 536 different quantizing intervals exist, whereas video systems typically have eight-bit systems having 256 quantizing intervals.

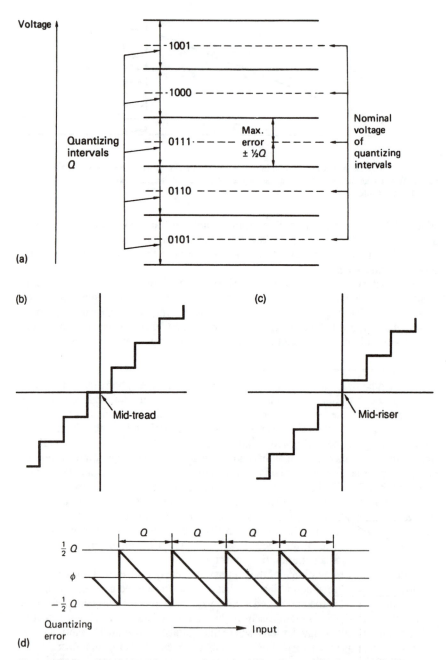

Figure 2.19 Quantizing assigns discrete numbers to variable voltages. All voltages within the same quantizing interval are assigned the same number which causes the DAC to produce the voltage at the centre of the intervals shown by the dashed lines at (a). This is the characteristic of the mid-tread quantizer shown at (b). An alternative system is the mid-riser system shown at (c). Here 0 volts analog falls between two codes and there is no code for zero. Such quantizing cannot be used prior to signal processing because the number is no longer proportional to the voltage. Quantizing error cannot exceed ±0.5Q as shown at (d)

2.7.9 Quantizing error

It is possible to draw a transfer function for such an ideal quantizer followed by an ideal DAC, and this is shown in Figure 2.19(b). A transfer function is simply a graph of the output with respect to the input. When the term linearity is used, this generally means the straightness of the transfer function. Linearity is a goal in audio and video, yet it will be seen that an ideal quantizer is anything but.

Figure 2.19(b) shows that the transfer function is somewhat like a staircase, and the voltage corresponding to audio muting or video blanking is half-way up a quantizing interval, or on the centre of a tread. This is the so-called mid-tread quantizer which is universally used in audio and video. Figure 2.19(c) shows the alternative mid-riser transfer function which causes difficulty because it does not have a code value at muting/blanking level and as a result the code value is not proportional to the signal voltage.

Quantizing causes a voltage error in the sample which is given by the difference between the actual staircase transfer function and the ideal straight line. This is shown in Figure 2.19(d) to be a sawtooth-like function periodic in Q. The amplitude cannot exceed $\pm 0.5Q$ unless the input is so large that clipping occurs.

In studying the transfer function it is better to avoid complicating matters with the aperture effect of the DAC. For this reason it is assumed here that output samples

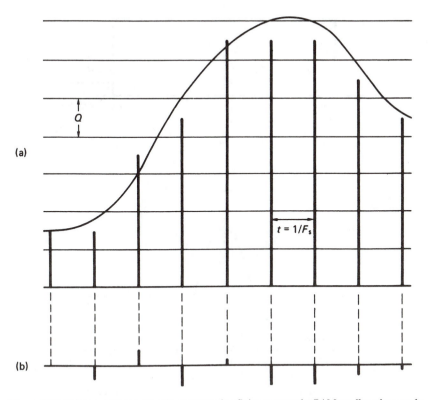

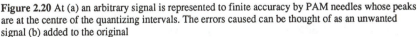

Figure 2.20 At (a) an arbitrary signal is represented to finite accuracy by PAM needles whose peaks are at the centre of the quantizing intervals. The errors caused can be thought of as an unwanted signal (b) added to the original

are of negligible duration. Then impulses from the DAC can be compared with the original analog waveform and the difference will be impulses representing the quantizing error waveform. As can be seen in Figure 2.20 the quantizing error waveform can be thought of as an unwanted signal which the quantizing process adds to the perfect original. As the transfer function is non-linear, ideal quantizing can cause distortion. As a result practical digital audio devices use non-ideal quantizers to achieve linearity. The quantizing error of an ideal quantizer is a complex function, and it has been researched in great depth[10]. It is not intended to go into such depth here. The characteristics of an ideal quantizer will only be pursued far enough to convince the reader that they cannot be used.

As the magnitude of the quantizing error is limited, its effect can be minimized by making the signal larger. This will require more quantizing intervals and more bits to express them. The number of quantizing intervals multiplied by their size gives the quantizing range of the convertor. A signal outside the range will be clipped. Clearly if clipping is avoided, the larger the signal the less will be the effect of the quantizing error.

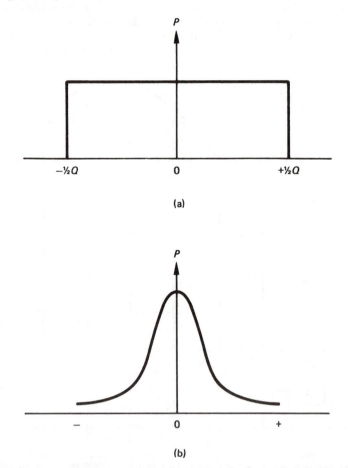

(a)

(b)

Figure 2.21 (a) The amplitude of a quantizing error needle will be from $-0.5Q$ to $+0.5Q$ with equal probability. (b) White noise in analog circuits generally has Gaussian amplitude distribution

Consider first the case where the input signal exercises the whole quantizing range and has a complex waveform. In audio this might be orchestral music; in video a bright, detailed contrast scene. In these cases successive samples will have widely varying numerical values and the quantizing error on a given sample will be independent of that on others. In this case the size of the quantizing error will be distributed with equal probability between the limits. Figure 2.21(a) shows the resultant uniform probability density. In this case the unwanted signal added by quantizing is an additive broadband noise uncorrelated with the signal, and it is appropriate in this case to call it quantizing noise. This is not quite the same as thermal noise which has a Gaussian probability shown in Figure 2.21(b). The subjective difference is slight. Treatments which then assume that quantizing error is *always* noise give results which are at variance with reality. Such approaches only work if the probability density of the quantizing error is uniform. Unfortunately at low levels, and particularly with pure or simple waveforms, this is simply not true.

At low levels, quantizing error ceases to be random, and becomes a function of the input waveform and the quantizing structure. Once an unwanted signal becomes a deterministic function of the wanted signal, it has to be classed as a distortion rather than a noise. We predicted a distortion because of the non-linearity or staircase nature of the transfer function. With a large signal, there are so many steps involved that we must stand well back, and a staircase with many steps appears to be a slope. With a small signal there are few steps and they can no longer be ignored.

The non-linearity of the transfer function results in distortion, which produces harmonics. Unfortunately these harmonics are generated *after* the anti-aliasing filter, and so any which exceed half the sampling rate will alias. Figure 2.22 shows how this results in anharmonic distortion in audio. These anharmonics result in spurious tones known as birdsinging. When the sampling rate is a multiple of the input frequency the result is harmonic distortion. This is shown in Figure 2.23. Where more than one frequency is present in the input, intermodulation distortion occurs, which is known as granulation.

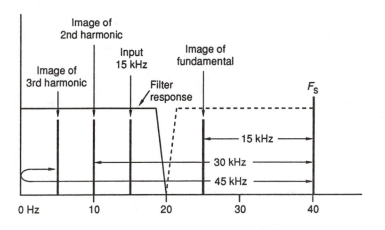

Figure 2.22 Quantizing produces distortion *after* the anti-aliasing filter, thus the distortion products will fold back to produce anharmonics in the audio band. Here the fundamental of 15 kHz produces 2nd and 3rd harmonic distortion at 30 and 45 kHz. This results in aliased products at 40–30 = 10 kHz and 40–45 = (–)5 kHz

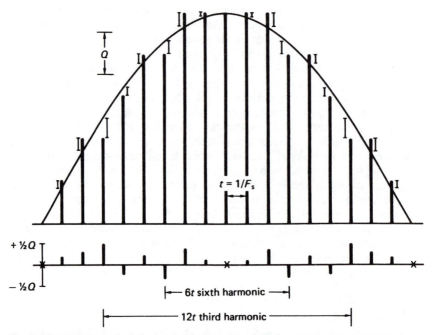

Figure 2.23 Mathematically derived quantizing error waveform for a sine wave sampled at a multiple of itself. The numerous autocorrelations between quantizing errors show that there are harmonics of the signal in the error, and that the error is not random, but deterministic

As the input signal is further reduced in level, it may remain within one quantizing interval. The output will be silent because the signal is now the quantizing error. In this condition, low frequency signals such as air-conditioning rumble can shift the input in and out of a quantizing interval so that the quantizing distortion comes and goes, resulting in noise modulation.

In video, quantizing error results in visible contouring on low key scenes or flat fields. Slowly changing brightness across the screen is replaced by areas of constant brightness separated by sudden steps.

Needless to say any one of the above effects would prevent the use of an ideal quantizer for high quality work. There is little point in studying the adverse effects further as they can be eliminated completely in practical equipment by the use of dither.

2.7.10 Dither

At high signal level, quantizing error is effectively noise. As the level falls, the quantizing error of an ideal quantizer becomes more strongly correlated with the signal and the result is distortion. If the quantizing error can be decorrelated from the input in some way, the system can remain linear. Dither performs the job of decorrelation by making the action of the quantizer unpredictable.

The first documented use of dither was in picture coding[11]. In this system, the noise added prior to quantizing was subtracted after reconversion to analog. This is known as subtractive dither. Although subsequent subtraction has some slight advantages[10], it suffers from practical drawbacks, since the original noise waveform must accompany the samples or must be synchronously re-created at the DAC. This is virtually impossible in a system where the signal may have been edited. Practical systems use non-subtractive dither where the dither signal is added prior to quantization and no subsequent attempt is made to remove it. The introduction of dither inevitably causes a slight reduction in the signal-to-noise ratio attainable, but this reduction is a small price to pay for the elimination of non-linearities. As linearity is an essential requirement for digital audio and video, the use of dither is equally essential and there is little point in deriving a signal-to-noise ratio for an undithered system as it has no relevance to real applications. Instead, a study of dither will be used to derive the dither amplitude necessary for linearity and freedom from artefacts, and the signal-to-noise ratio available should follow from that.

The ideal (noiseless) quantizer of Figure 2.19 has fixed quantizing intervals and must always produce the same quantizing error from the same signal. In Figure 2.24 it can be seen that an ideal quantizer can be dithered by linearly adding a controlled level of noise either to the input signal or to the reference voltage which is used to derive the quantizing intervals. There are several ways of considering how dither works, all of which are valid. The addition of dither means that successive samples effectively find the quantizing intervals in different places on the voltage scale. The

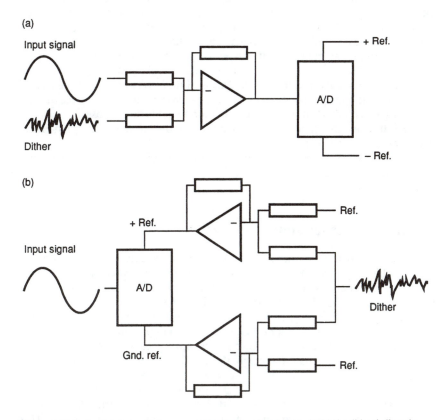

Figure 2.24 Dither can be applied to a quantizer in one of two ways. At (a) the dither is linearly added to the analog input signal whereas at (b) it is added to the reference voltages of the quantizer

quantizing error becomes a function of the dither, rather than just a function of the input signal. The quantizing error is not eliminated, but the subjectively unacceptable distortion is converted into broadband noise which is more benign.

That noise may be also frequency shaped to take account of the frequency shaping of the ear and eye responses.

An alternative way of looking at dither is to consider the situation where a low level input signal is changing slowly within a quantizing interval. Without dither, the same numerical code results, and the variations within the interval are lost. Dither has the effect of forcing the quantizer to switch between two or more states. The higher the voltage of the input signal within the interval, the more probable it becomes that the output code will take on a higher value. The lower the input voltage within the interval, the more probable it is that the output code will take the lower value. The dither has resulted in a form of duty cycle modulation, and the resolution of the system has been extended indefinitely instead of being limited by the size of the steps.

Dither can also be understood by considering the effect it has on the transfer function of the quantizer. This is normally a perfect staircase, but in the presence of dither it is smeared horizontally until with a certain minimum amplitude the average transfer function becomes straight.

The characteristics of the noise used are rather important for optimal performance, although many sub-optimal but nevertheless effective systems are in use. The main parameters of interest are the peak-to-peak amplitude, and the probability distribution of the amplitude. Triangular probability works best and this can be obtained by summing the output of two uniform probability processes.

The use of dither invalidates the conventional calculations of signal-to-noise ratio available for a given word length. This is of little consequence as the rule of thumb that multiplying the number of bits in the word length by 6 dB gives the S/N ratio gives a result that will be close enough for all practical purposes.

The technique can also be used to convert signals quantized with more intervals to a lower number of intervals.

It has only been possible to introduce the principles of conversion of audio and video signals here. For more details of the operation of convertors the reader is referred elsewhere[12,13].

2.8 Binary codes for audio

For audio use, the prime purpose of binary numbers is to express the values of the samples which represent the original analog sound-pressure waveform. There will be a fixed number of bits in the sample, which determines the number range. In a 16 bit code there are 65 536 different numbers. Each number represents a different analog signal voltage, and care must be taken during conversion to ensure that the signal does not go outside the convertor range, or it will be clipped. In Figure 2.25, it will be seen that in a unipolar system, the number range goes from 0000 hex, which represents the largest negative voltage, through 7FFF hex, which represents the smallest negative voltage, through 8000 hex, which represents the smallest positive voltage, to FFFF hex, which represents the largest positive voltage. Effectively, the number range of the convertor has been shifted so that positive and negative voltages in a bipolar signal can be expressed by binary numbers which are only positive. This approach is called offset binary, and is perfectly acceptable where the signal has been

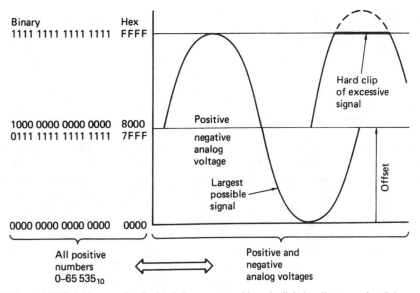

Figure 2.25 Offset binary coding is simple but causes problems in digital audio processing. It is seldom used

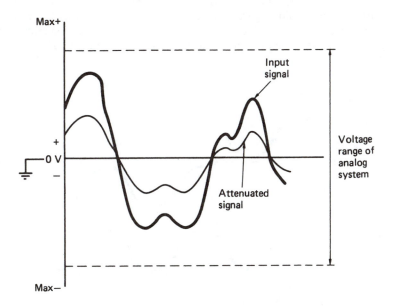

Figure 2.26 Attenuation of an audio signal takes place with respect to midrange

digitized only for recording or transmission from one place to another, after which it will be converted back to analog. Under these conditions it is not necessary for the quantizing steps to be uniform, provided both ADC and DAC are constructed to the same standard. In practice, it is the requirements of signal processing in the digital domain which make both non-uniform quantizing and offset binary unsuitable.

Figure 2.26 shows that an audio signal voltage is referred to midrange. The level of the signal is measured by how far the waveform deviates from midrange, and attenuation, gain and mixing all take place around midrange. It is necessary to add sample values from two or more different sources to perform the mixing function, and adding circuits assume that all bits represent the same quantizing interval so that the sum of two sample values will represent the sum of the two original analog voltages. In non-uniform quantizing this is not the case, and such signals cannot readily be processed. If two offset binary sample streams are added together in an attempt to perform digital mixing, the result will be an offset which may lead to an overflow. Similarly, if an attempt is made to attenuate by, say, 6 dB by dividing all of the sample values by two, Figure 2.27 shows that a further offset results. The problem is that offset binary is referred to one end of the range. What is needed is a numbering system which operates symmetrically about the centre of the range.

In the two's complement system, which has this property, the upper half of the pure binary number range has been defined to represent negative quantities. If a pure binary counter is constantly incremented and allowed to overflow, it will produce all the numbers in the range permitted by the number of available bits, and these are shown for a 4 bit example drawn around the circle in Figure 2.28. In two's complement, however, the number range this represents does not start at zero, but starts on the opposite side of the circle. Zero is midrange, and all numbers with the most significant bit set are considered negative. This system allows two sample values to be added, where the result is referred to the system midrange; this is analogous to adding analog signals in an operational amplifier. A further asset of two's complement notation is that binary subtraction can be performed using only adding logic. The two's complement is added to perform a subtraction. This permits

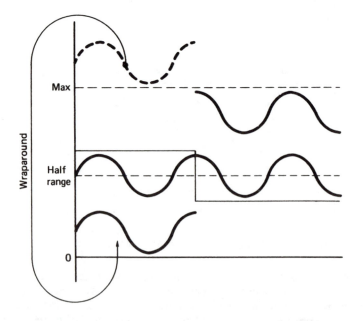

Figure 2.27 If two pure binary data streams are added to simulate mixing, offset or overflow will result

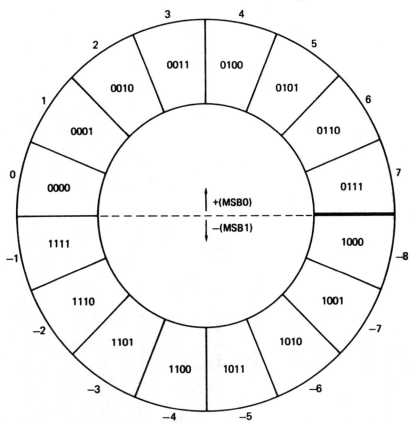

Figure 2.28 In this example of a 4 bit two's complement code, the number range is from –8 to +7. Note that the MSB determines polarity

(a) **Conversion to two s complement from binary**
Positive numbers: add leading zeros to determine sign bit.
Example: $101_2 = 5_{10} = 0101_{2C}$

Negative numbers: add leading zeros to final
wordlength; invert all bits; add one.
Example 1: $11_2 = 3_{10} \rightarrow 0011 \rightarrow 1100 \rightarrow 1101_{2C} = -3$

                      add   add
                    leading   invert   1
                    zeros

Example 2: $100_2 = 4_{10} \rightarrow 0100 \rightarrow 1011 \rightarrow 1100_{2C} = -4$

(b) **Conversion to binary from two's complement**
If MSB = 1 (Negative Number), invert all bits; add one.

Example 1: $1001 \rightarrow 0110 \rightarrow 0111 = -7_{10}$

             invert   add
                     1

Example 2: $1110 \rightarrow 0001 \rightarrow 0010 = -2_{10}$

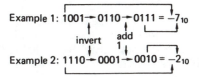

Figure 2.29 (a) Two's complement conversion from binary. (b) Binary conversion from two's complement

(c)   4   4   0100
     −6 ≡ + (−6) ≡ + 1010
     ─────   ─────   ───────
     −2   −2   1110

     −8   1000
     +3 ≡ 0011
     ───   ──────
     −5   1011

     −3   1101
     +6 ≡ 0110
     ──   ──────
      3   0011
            ᴜᴜ
            C C

Figure 2.29(c) Some examples of two's complement arithmetic

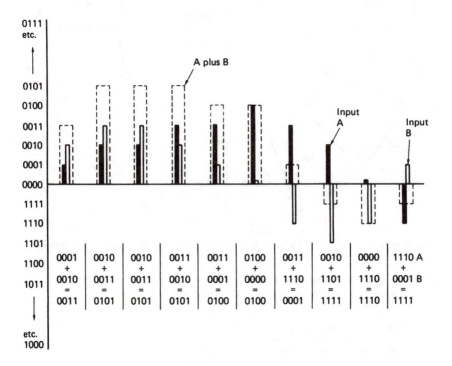

Figure 2.29(d) Using two's complement arithmetic, single values from two waveforms are added together with respect to midrange to give a correct mixing function

a significant saving in hardware complexity, since only carry logic is necessary and no borrow mechanism need be supported.

For these reasons, two's complement notation is in virtually universal use in digital audio processing, and is accordingly adopted by all the major digital recording formats, and consequently it is specified for the AES/EBU digital audio interface as well as for all of the other audio interfaces described in this book.

Fortunately the process of conversion to two's complement is simple. The signed binary number which is to be expressed as a negative number is written down with

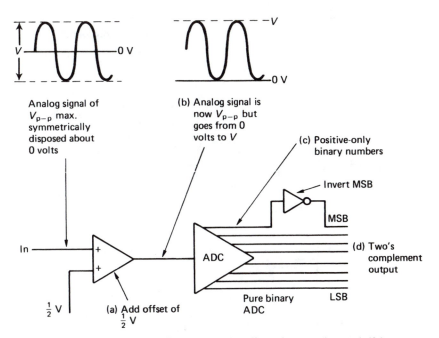

Figure 2.30 A two's complement ADC. At (a) an analog offset voltage equal to one-half the quantizing range is added to the bipolar analog signal in order to make it unipolar as at (b). The ADC produces positive only numbers at (c), but the MSB is then inverted at (d) to give a two's complement output

leading zeros if necessary to occupy the wordlength of the system. All bits are then inverted, to form the one's complement, and one is added. To return to signed binary, if the most significant bit of the two's complement number is false, no action need be taken. If the most significant bit is set, the sign is negative, all bits are inverted, and one is added. Figure 2.29 shows some examples of conversion to and from two's complement, and illustrates how adding two's complement samples simulates the mixing process.

Reference to Figure 2.28 will show that inverting the MSB results in a jump to the diametrically opposite code which is the equivalent of an offset of half full scale. Thus it is possible to obtain two's complement coded data by adding an offset of half full scale to the analog input of a convertor and then inverting the MSB as shown in Figure 2.30.

2.9 Binary codes for video

There are a wide variety of waveform types encountered in video, and all of these can be successfully digitized. All that is needed is for the quantizing range to be a little greater than the useful voltage range of the waveform.

Composite video contains an embedded subcarrier, which has meaningful levels above white and below black. The quantizing range has to be extended to embrace the total possible excursion of luminance plus subcarrier. The resultant range is so close to the overall range of the signal that, in composite working, the whole signal,

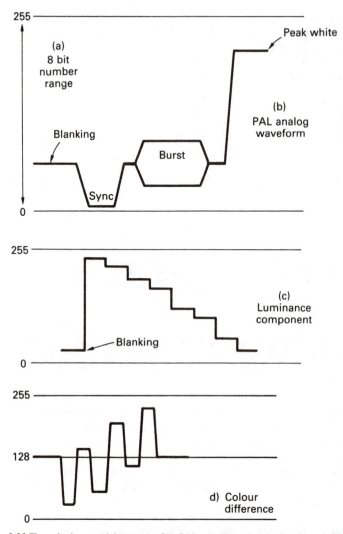

Figure 2.31 The unipolar quantizing range of an 8 bit pure binary system is shown at (a). The analog input must be shifted to fit into the quantizing range, as shown for PAL at (b). In component, sync pulses are not digitized, so the quantizing intervals can be smaller as at (c). An offset of half scale is used for colour difference signals (d)

including syncs, is made to fit into the quantizing range as Figure 2.31(b) shows. This is particularly useful in PAL because the sampling points are locked to subcarrier, and have a complex relationship to sync. Clearly black level no longer corresponds to digital zero. In eight-bit PAL it is 64_{10}. It is as if a constant of 64 had been added to every sample, and gives rise to the term offset binary.

In the luminance component signal, the useful waveform is unipolar and can only vary between black and white, since the syncs carry no information which cannot be recreated. Only the active line is coded and the quantizing range is optimized to suit the gamut of unblanked luminance as shown in Figure 2.31(c).

Chroma and colour difference signals are bipolar and can have values above and below blanking. In this they have the same characteristics as audio and require to be *processed* using two's complement coding as described in the previous section. However, two's complement is not used for digital video *interfacing* as the codes of all ones and all zeros are reserved for synchronizing and in two's complement these would appear in the centre of the quantizing range. Instead digital video interfaces use an offset of half full scale to shift blanking level to the centre of the scale as shown in Figure 2.31(d). All ones and all zeros are then at the ends of the scale. Such an offset binary signal can easily be converted to two's complement by inverting the MSB as shown in Figure 2.30. If the quantizing range were set to exactly the video signal range or gamut, a slightly excessive gain somewhere in the analog input path would cause clipping. In practice the quantizing range will be a little greater than the nominal signal range and as a result even the unipolar luminance signal has an offset blanking level. A video interface format must specify the numerical values which result from reference analog levels so that all transmissions will result in identical signals at all receiving devices.

2.10 Requantizing and digital dither

The advanced ADC technology which is now available allows 18 and 20 bit resolution to be obtained in audio, with perhaps more in the future. The situation then arises that an existing 16 bit device such as a digital recorder needs to be connected to the output of an ADC with greater wordlength. In a similar fashion digital video equipment has recently moved up from eight to ten bit working with the result that ten-bit signals are presented to eight-bit devices. In both cases the words need to be shortened in some way.

When a sample value is attenuated, the extra low-order bits which come into existence below the radix point preserve the resolution of the signal and the dither in the least significant bit(s) which linearizes the system. The same word extension will occur in any process involving multiplication, such as mixing or digital filtering. It will subsequently be necessary to shorten the wordlength. Clearly the high order bits cannot be discarded in two's complement as this would cause clipping of positive half cycles and a level shift on negative half cycles due to the loss of the sign bit. Low-order bits must be removed instead. Even if the original conversion was correctly dithered, the random element in the low-order bits will now be some way below the end of the intended word. If the word is simply truncated by discarding the unwanted low-order bits or rounded to the nearest integer the linearizing effect of the original dither will be lost. In audio the result will be low-level distortion; in video the result will be contouring.

Shortening the wordlength of a sample reduces the number of quantizing intervals available without changing the signal amplitude. As Figure 2.32 shows, the quantizing intervals become larger and the original signal is *requantized* with the new interval structure. This will introduce requantizing distortion having the same characteristics as quantizing distortion in an ADC. It then is obvious that when shortening the wordlength of a 20 bit convertor to 16 bits, the four low-order bits must be removed in a way that displays the same overall quantizing structure as if the original convertor had been only of 16 bit wordlength. It will be seen from Figure 2.32 that truncation cannot be used because it does not meet the above requirement but results in signal dependent offsets because it always rounds in the same

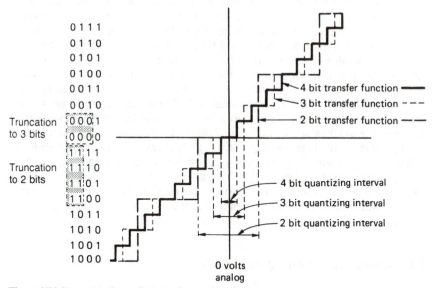

Figure 2.32 Shortening the wordlength of a sample reduces the number of codes which can describe the voltage of the waveform. This makes the quantizing steps bigger, hence the term requantizing. It can be seen that simple truncation or omission of the bits does not give analogous behaviour. Rounding is necessary to give the same result as if the larger steps had been used in the original conversion

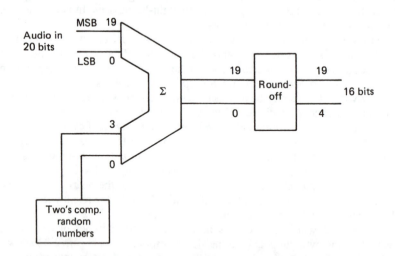

Figure 2.33 In a simple digital dithering system, two's complement values from a random number generator are added to low-order bits of the input. The dithered values are then rounded up or down according to the value of the bits to be removed. The dither linearizes the requantizing

direction. Proper numerical rounding is essential in audio and video applications. Rounding in two's complement is a little more complex than in pure binary.

Requantizing by numerical rounding accurately simulates analog quantizing to the new interval size. Unfortunately the 20 bit convertor will have a dither amplitude appropriate to quantizing intervals one-sixteenth the size of a 16 bit unit and the result will be highly non-linear.

In practice, the wordlength of samples must be shortened in such a way that the requantizing error is converted to noise rather than distortion. One technique which meets this requirement is to use digital dithering[14] prior to rounding. This is directly equivalent to the analog dithering in an ADC.

Digital dither is a pseudo-random sequence of numbers. If it is required to simulate the analog dither signal of Figure 2.24, then it is obvious that the noise must be bipolar so that it can have an average voltage of zero. Two's complement coding must be used for the dither values to obtain this characteristic.

Figure 2.33 shows a simple digital dithering system (i.e. one without noise shaping) for shortening sample wordlength. The output of a two's complement pseudo-random sequence generator of appropriate wordlength is added to input samples prior to rounding. The most significant of the bits to be discarded is examined in order to determine whether the bits to be removed sum to more or less than half a quantizing interval. The dithered sample is either rounded down, i.e. the unwanted bits are simply discarded, or rounded up, i.e. the unwanted bits are discarded but 1 is added to the value of the new short word. The rounding process is no longer deterministic because of the added dither which provides a linearizing random component.

If this process is compared with that of Figure 2.24 it will be seen that the principles of analog and digital dither are identical; the processes simply take place in different domains using numbers which are rounded or voltages which are quantized as appropriate. In fact quantization of an analog dithered waveform is identical to the hypothetical case of rounding after bipolar digital dither where the number of bits to be removed is infinite, and remains identical for practical purposes when as few as eight bits are to be removed. The probability density of the pseudo-random sequence is important. Vanderkooy and Lipshitz[14] found that uniform probability density produced noise modulation, in which the amplitude of the random component varies as a function of the amplitude of the samples. A triangular probability density function obtained by adding together two pseudo-random sequences eliminated the noise modulation to yield a signal-independent white-noise component in the least significant bit. It is vital that such steps are taken when sample wordlength is to be reduced. More recently, sequences yielding signal-independent noise with a weighted frequency component have been used to improve the subjective effects of the noise component leading to a perceived reduction in noise.

References

1   BETTS, J. A. (1970) *Signal Processing Modulation and Noise*, Ch. 6, Hodder and Stoughton, Sevenoaks
2   MEYER, J. (1984) Time correction of anti-aliasing filters used in digital audio systems. *J. Audio Eng. Soc.*, vol. 32, pp. 132–137
3   BLESSER, B. (1983) Advanced A/D conversion and filtering: data conversion. In *Digital Audio*, ed. B. A. Blesser, B. Locanthi and T. G. Stockham Jr, pp. 37–53, Audio Engineering Society, New York
4   LAGADEC, R., WEISS, D. and GREUTMANN, R. (1980) High-quality analog filters for digital audio. Presented at the *67th Audio Engineering Society Convention* (New York), preprint 1707(B-4)
5   HSU, S. (1986) The Kell factor: past and present. *SMPTE Journal*, vol. 95, pp. 206–214
6   JESTY, L. C. (1958) The relationship between picture size, viewing distance and picture quality. *Proc. IEE*, vol. 105B, pp. 425–439

7   ISHIDA, Y. *et al.* (1979) A PCM digital audio processor for home use VTRs. Presented at *64th Audio Engineering Society Convention* (New York), preprint 1528

8   WATKINSON, J. R. (1994) *The Digital Video Tape Recorder*, Focal Press, Oxford

9   AES (1984) AES recommended practice for professional digital audio applications employing pulse code modulation: preferred sampling frequencies. AES5-1984 (ANSI S4.28-1984), *J. Audio Eng. Soc.*, vol. 32, pp. 781–785

10 LIPSHITZ, S. P. *et al.* (1992) Quantization and dither: a theoretical survey. *J. Audio Eng. Soc.*, vol. 40, pp. 355–375

11 ROBERTS, L. G. (1962) Picture coding using pseudo-random noise. *IRE Trans. Inform. Theory*, vol. IT-8, pp. 145–154

12 WATKINSON, J. R. (1994) *The Art of Digital Audio*, Second Edn, Focal Press, Oxford

13 WATKINSON, J. R. (1994) *The Art of Digital Video*, Second Edn, Focal Press, Oxford

14 VANDERKOOY, J. and LIPSHITZ, S. P. (1986) Digital dither. Presented at the *81st Audio Engineering Society Convention* (Los Angeles), preprint 2412(C-8)

Chapter 3

Digital transmission

Although the physics of the transmission process is unaffected by the meaning attributed to signals, the techniques used in digital transmission are rather different from those used with analog signals, although often the same phenomenon shows up in a different guise. In this chapter the fundamentals of digital transmission are introduced along with descriptions of the coding and error detection techniques used in practical applications.

The subject of error detection is almost always described in mathematical terms by specialists for the benefit of other specialists. Such mathematical approaches are quite inappropriate for a proper understanding of the concepts of error detection and only become necessary to analyse the quantitative behaviour of a system. The mathematics behind error detection is the same as that behind many modern transmission systems such as SDI (serial digital interface) and NICAM. The descriptions here will use the minimum possible amount of mathematics, and it will then be seen that the subject is, in fact, quite straightforward.

3.1 Data transmission principles

3.1.1 Introduction to the channel

In analog transmission, the characteristics of the channel affect the signal directly, whereas by expressing a signal in binary numerical form by sampling and quantizing, the quality can be made independent of the channel. The dynamic range required in the programme material no longer directly decides the dynamic range of the channel. In digital circuitry there is a great deal of noise immunity because the signal can only have two states, which are widely separated compared with the amplitude of noise. In digital transmission the signal may originate with discrete states which change at discrete times, but the channel will treat it as an analog waveform. Various loss mechanisms will attenuate the amplitude of the signal. These attenuations will not be the same at all frequencies. Noise will be picked up in the channel as a result of stray electric fields or magnetic induction. As a result the voltage received at the end of the channel will have an infinitely varying state along with a degree of uncertainty due to the noise. Different frequencies can propagate at different speeds in the channel; this is the phenomenon of group delay. An alternative way of considering group delay is that there will be frequency-dependent phase shifts in the signal and these will result in uncertainty in the timing of pulses.

So it is not the transmission channel which is digital; instead the term describes the way in which the received signals are interpreted. When the receiver makes discrete decisions from the input waveform the uncertainties in voltage and time need to be rejected. The technique of channel coding is one where transmitted waveforms are restricted to those which still allow the receiver to make discrete decisions despite the degradations caused by the analog nature of the channel.

3.1.2 Sensitivity of message to error

Before attempting to specify any piece of equipment, it is necessary to quantify the problems to be overcome and how effectively they need to be overcome. For a digital transmission system the causes of errors must be studied to quantify the problem, and the sensitivity of the destination to errors must be assessed. In audio and video the sensitivity to errors must be subjective. In both, the effect of a single bit in error depends upon the significance of the bit. If the least significant bit of a sample is wrong, the chances are that the effect will be lost in the noise. Conversely, if a high-order bit is in error, a massive transient will be added to the waveform. In audio this is almost certain to be discernible; in video it is less certain but still likely. If the error rate demanded by the destination cannot be met by the unaided channel, some form of error handling will be necessary.

3.1.3 Error mechanisms

Errors in data transmission tend to fall into two categories. There are large isolated corruptions, called error bursts, where numerous bits are corrupted all together in an area which is otherwise error free, and there are random errors affecting single bits or symbols. Burst errors may be due, for example, to lightning discharge interfering with a radio transmission whereas random errors are due to continuous processes such as noise. Whatever the mechanism, the result will be that the received data will not be exactly the same as that sent.

3.1.4 Noise

Noise in the channel causes uncertainty about the voltage of the reproduced signal. Noise is statistical, and can have any amplitude, except that higher amplitudes occur with decreasing probability. When the noise exceeds a certain amplitude, it can corrupt data. In digital recorders, it is possible to allow greater noise and more frequent errors because an error correction system is generally used. The noise tolerance gained by the use of error correction allows higher recording densities to be obtained. However, error correction processes cause delay, and whilst this is acceptable in, for example, the replay of a tape, it may not be acceptable in an audio or video interface because of the consequent timing problems. Interfaces covered in this book generally do not use error correction so that they cause minimal signal delay. They will, however, often use some form of error detection so that an indication of marginal operation is available.

The lack of error correction requires that noise on interfaces is kept moderate. On a properly installed interface noise should be substantially lower in amplitude than the signal. Noise then has a slightly different effect, which can be seen in Figure 3.1. When noise is present on a sloping signal edge, it can change the time at which the slicer judges the transition to have occurred. This is sometimes known as noise jitter.

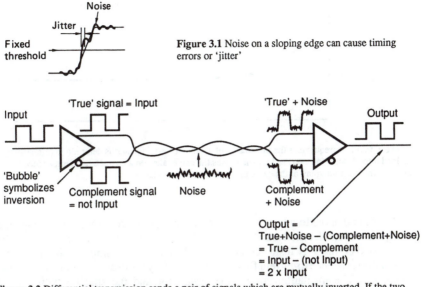

Figure 3.1 Noise on a sloping edge can cause timing errors or 'jitter'

Figure 3.2 Differential transmission sends a pair of signals which are mutually inverted. If the two conductors are twisted together the noise picked up will be substantially the same and can be cancelled in the receiver which subtracts the two signals to produce a unipolar output. In many cases the receiver is a comparator which slices and subtracts simultaneously

Differential signals are often used in interfacing to reduce sensitivity to noise. Figure 3.2 shows that in differential working, a pair of conductors is used for each signal. These are driven in anti-phase such that when the voltage on one falls, the voltage on the other rises. The conductors are often twisted together as this ensures that both pick up as nearly as possible the same external noise. Twisting also keeps the conductor spacing and hence the impedance constant. The receiver subtracts the anti-phase signal from the in-phase signal. As a result the noise is cancelled but the signals add.

3.1.5 Transmission modes

Transmission can be by electrical conductors, radio or optical fibre. Although these appear to be completely different, they are in fact just different examples of electromagnetic energy travelling from one place to another. If the energy is made to vary in some way, information can be carried.

Even today electromagnetism is not fully understood, but we have sufficiently good models based on experimental results so that practical equipment can be made. It is not actually necessary to fully understand a process in order to harness it; it is only necessary to be able to reliably predict what will happen in given circumstances.

Electromagnetic energy propagates in a manner which is a function of frequency, and our lack of understanding requires it to be considered as electrons, waves or photons so that we can predict its behaviour in given circumstances.

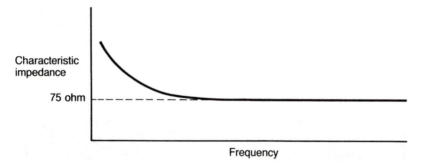

Figure 3.3 At low frequencies the impedance of a cable is very high and is determined primarily by the load impedance. However, as frequency rises the cable takes on a characteristic impedance resulting from the interplay of the cable inductance and capacitance

At DC and at the low frequencies used for power distribution, electromagnetic energy is called electricity and it is remarkably aimless stuff which needs to be transported completely inside conductors. It has to have a complete circuit to flow in, and the resistance to current flow is determined by the cross-sectional area of the conductor. The insulation around the conductor and the spacing between the conductors has no effect on the ability of the conductor to pass current. At DC an inductor appears to be a short circuit, and a capacitor appears to be an open circuit.

As frequency rises, resistance is exchanged for impedance. Inductors display increasing impedance with frequency, capacitors show falling impedance. Electromagnetic energy becomes increasingly desperate to leave the conductor. The first symptom is that the current flows only in the outside layer of the conductor, effectively causing the resistance to rise. This is the skin effect and gives rise to such techniques as Litz wire which has as much surface area as possible per unit cross-section, and to silver-plated conductors in which the surface has lower resistivity than the interior.

As the energy is starting to leave the conductors, the characteristics of the space between them become important. This determines the impedance.

Figure 3.3 shows how the impedance of a cable changes with frequency. At DC and low frequencies it is effectively an open circuit. As frequency rises eventually the dimensions and the dielectric constant of the insulating material dominate. At analog audio frequencies, cabling has effectively infinite impedance. The common 600 ohm analog audio interface only has such an impedance because of the load which the cable drives.

A change of impedance causes reflections in the energy flow and some of it heads back in the opposite direction. Constant impedance cables with fixed conductor spacing are necessary, and these must be suitably terminated.

As frequency rises still further, the energy travels less in the conductors and more in the insulation between them, and their composition becomes important and they begin to be called dielectrics. A poor dielectric like PVC absorbs high frequency energy and attenuates the signal. The response of a cable becomes proportional to the reciprocal of the square root of the frequency. At high frequencies, so-called low-loss dielectrics such as PTFE are used, and one way of achieving low loss is to incorporate as much air in the dielectric as possible by making it in the form of a foam or extruding it with voids.

Further rise in frequency causes the energy to start behaving more like waves and less like electron movement. As the wavelength falls it becomes increasingly directional. The transmission line becomes a waveguide and microwaves are sufficiently directional that they can keep going without any conductor at all. Microwaves are simply low frequency radiant heat, which is itself low frequency light. All three are reflected well by electrical conductors, and can be refracted at the boundary between media having different propagation speeds. A waveguide is the microwave equivalent of an optical fibre.

This frequency-dependent behaviour is the most important factor in deciding how best to harness electromagnetic energy flow for information transmission. It is obvious that the higher the frequency, the greater the possible information rate, but in general, losses increase with frequency, and flat frequency response is elusive. The best that can be managed is that over a narrow band of frequencies, the response can be made reasonably constant with the help of equalization. Unfortunately raw data when serialized have an unconstrained spectrum. Runs of identical bits can produce frequencies much lower than the bit rate would suggest. One of the essential steps in a transmission system is to modify the spectrum of the data into something more suitable.

At moderate bit rates, say a few megabits per second, and with moderate cable lengths, say a few metres, the dominant effect will be the capacitance of the cable due to the geometry of the space between the conductors and the dielectric between. The capacitance behaves under these conditions as if it were a single capacitor connected across the signal. Figure 3.4 shows the equivalent circuit.

The effect of the series source resistance and the parallel capacitance is that signal edges or transitions are turned into exponential curves as the capacitance is effectively being charged and discharged through the source impedance. This effect can be observed on the AES/EBU interface with short cables. Although the position where the edges cross the centreline is displaced, the signal eventually reaches the same amplitude as it would at DC.

As cable length increases, the capacitance can no longer be lumped as if it were a single unit; it has to be regarded as being distributed along the cable. With rising frequency, the cable inductance also becomes significant, and it too is distributed

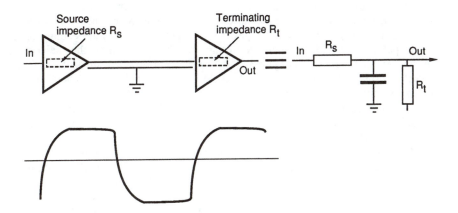

Figure 3.4 With a short cable, the capacitance between the conductors can be lumped as if it were a discrete component. The effect of the parallel capacitor is to slope off the edges of the signal

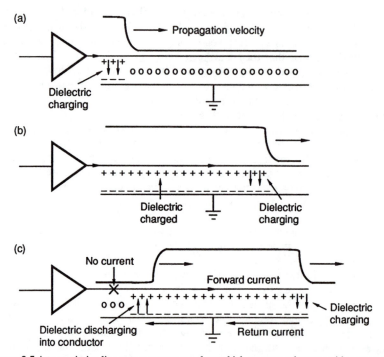

Figure 3.5 A transmission line conveys energy packets which appear to alternate with respect to the dielectric. At (a) the driver launches a pulse which charges the dielectric at the beginning of the line. As it propagates the dielectric is charged further along as in (b). When the driver ends the pulse, the charged dielectric discharges into the line. A current loop is formed where the current in the return loop flows in the opposite direction to the current in the 'hot' wire

The cable must now be considered a transmission line and pulses travel down it as current loops which roll along as shown in Figure 3.5. If the pulse is positive, as it is launched along the line, it will charge the dielectric locally as in Figure 3.5(a). As the pulse moves along, it will continue to charge the local dielectric as in Figure 3.5(b). When the driver finishes the pulse, the trailing edge of the pulse follows the leading edge along the line. The voltage of the dielectric charged by the leading edge of the pulse is now higher than the voltage on the line, and so the dielectric discharges into the line as in Figure 3.5(c). The current flows forward as it is in fact the same current which is flowing into the dielectric at the leading edge. There is thus a loop of current rolling down the line flowing forward in the 'hot' wire and backwards in the return. The analogy with the tracks of a Caterpillar tractor is quite good. Individual plates in the track find themselves being lowered to the ground at the front and raised again at the back.

The constant to-ing and fro-ing of charge in the dielectric results in dielectric loss of signal energy. Dielectric loss increases with frequency and so a long transmission line acts as a filter. Thus the term 'low-loss' cable refers primarily to the kind of dielectric used.

Transmission lines which transport energy in this way have a characteristic impedance caused by the interplay of the inductance along the conductors with the parallel capacitance. One consequence of that transmission mode is that correct termination or matching is required between the line and both the driver and the receiver. When a line is correctly matched, the rolling energy rolls straight out of the line into the load and the maximum energy is available. If the impedance presented by the load is incorrect, there will be reflections from the mismatch. An open circuit will reflect all of the energy back in the same polarity as the original, whereas a short circuit will reflect all of the energy back in the opposite polarity. Thus impedances above or below the correct value will have a tendency toward reflections whose magnitude depends upon the degree of mismatch and whose polarity depends upon whether the load is too high or too low. In practice it is the need to avoid reflections

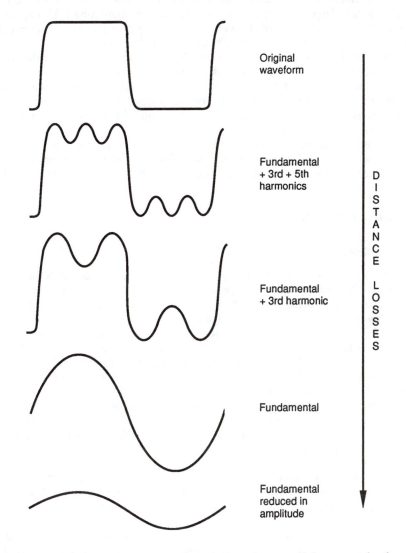

Figure 3.6 A signal may be square at the transmitter, but losses increase with frequency and as the signal propagates, more of the harmonics are lost until only the fundamental remains. The amplitude of the fundamental then falls with further distance

which is the most important reason to correctly terminate. The return loss parameter is a measure of the degree of impedance matching obtained.

Reflections at impedance mismatches have practical applications; electricity companies inject high frequency pulses into faulty cables and the time taken until the reflection from the break or short returns can be used to locate the source of damage. The same technique can be used to find wiring breaks in large studio complexes.

In professional video equipment both analog and digital video cables have an impedance of 75 ohm, but oddly employ 50 ohm BNC connectors. The original BNC connector was available in 50 ohm and 75 ohm versions, but the 50 ohm version had a larger centre pin which was found to be more robust. This apparent mismatch does not cause any difficulty as the quality of termination is determined primarily by the impedance of the circuitry behind the connector.

A perfectly square pulse contains an indefinite series of harmonics, but the higher ones suffer progressively more loss. A square pulse at the driver becomes less and less square with distance, as Figure 3.6 shows. The harmonics are progressively lost until in the extreme case all that is left is the fundamental. A transmitted square wave is received as a sine wave. Fortunately data can still be recovered from the fundamental signal component.

Once all the harmonics have been lost, further losses cause the amplitude of the fundamental to fall. The effect worsens with distance and it is necessary to ensure that data recovery is still possible from a signal of unpredictable level.

3.2 Data recovery

3.2.1 Slicing

Figure 3.7 shows that an initial step in data recovery at the receiver is the slicing process. The received signal voltage is compared with the midway voltage, known as the baseline or slicing level, using a comparator. If the signal voltage is above the baseline, the comparator outputs a high level; if below, a low level results.

In Figure 3.7(a) the signal to be sliced is a sine wave which is the fundamental which remains from a transmitted square wave. As the amplitude falls, the waveform from the slicer remains essentially the same. However, in Figure 3.7(b) the transmitted waveform has an uneven duty cycle. The DC component, or average level, of the signal is received with high amplitude, but the pulse amplitude falls as the pulse gets shorter. Eventually the waveform cannot be sliced.

In Figure 3.7(c) the opposite duty cycle is shown. The signal level drifts to the opposite polarity and once more slicing is impossible. The phenomenon is called baseline wander and will be observed with any signal whose average voltage is not the same as the slicing level.

In Figure 3.7(d) it will be seen that if the transmitted waveform has a relatively constant average voltage, slicing remains possible up to high frequencies even in the presence of serious loss, because the received waveform remains symmetrical about the baseline.

It is clearly not possible to simply serialize data in a shift register for so-called direct transmission, because successful slicing can only be obtained if the number of ones is equal to the number of zeros; there is little chance of this happening consistently with real data.

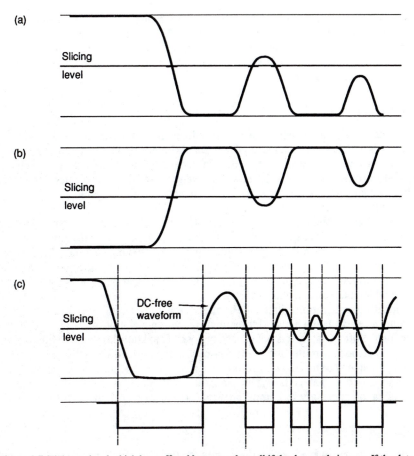

Figure 3.7 Slicing a signal which has suffered losses works well if the duty cycle is even. If the duty cycle is uneven, as at (a), timing errors will become worse until slicing fails. With the opposite duty cycle, the slicing fails in the opposite direction as at (b). If, however, the signal is DC free, correct slicing can continue even in the presence of serious losses, as (c) shows

Instead, a modulation code or channel code is necessary. This converts the data into a waveform which is DC-free or nearly so for the purpose of transmission.

3.2.2 Clocking

In digital circuitry the signals are generally accompanied by a separate clock signal. A further problem with transmission is that provision of a separate clock is not feasible except over short distances. A separate clock line would not only raise cost, but is impractical because at high frequency it is virtually impossible to ensure that the clock cable propagates signals at the same speed as the data cable. Propagation speed differences are called skew.

In certain serial digital video interfaces, the bit rate is 270 megabits per second. Each bit lasts for 3.7 ns. Assuming a 100 metre cable, there will be around 70 bits actually in the cable. If there were a separate clock, there would be roughly the same number of cycles in the clock cable, but no guarantee that they would propagate at the same speed or arrive in phase. The solution is to use a self-clocking waveform. A further essential function of the channel code is to create a waveform which carries its own clock content. Clearly, direct transmission does not have this characteristic. If all the data bits are the same, for example all zeros, there is no clock when they are serialized.

3.2.3 Equalization

The characteristics of most transmission lines are that signal loss occurs which increases with frequency. This has the effect of slowing down rise times and thereby sloping off edges. If a signal with sloping edges is sliced, the time at which the waveform crosses the slicing level will be changed, and this causes jitter. Figure 3.1 showed that slicing a sloping waveform in the presence of baseline wander causes more jitter.

On a long cable, high frequency roll-off can cause sufficient jitter to move a transition into an adjacent bit period. This is called intersymbol interference and the effect becomes worse in signals which have greater asymmetry, i.e. short pulses alternating with long ones. The effect can be reduced by the application of equalization, which is typically a high frequency boost, and by choosing a channel code which has restricted asymmetry.

Ideally, equalization ought to be applied at the transmitting end of a cable as this would result in better noise figures than if the equalization were applied at the receiver. Unfortunately it is not generally possible to know how much equalization to apply at the transmitter and so in practice equalization is done at the receiver. The adjustment may be manual or, in more recent equipment, automatic. If the spectrum of the coded signal is fairly constant, it is possible to assess the degree of equalization necessary by comparing the signal level at two different frequencies.

3.2.4 Jitter rejection

The squared-up waveform at the output of the slicer will be a replica of the transmitted waveform, except for the addition of jitter or time uncertainty in the position of the edges due to noise, baseline wander, intersymbol interference and imperfect equalization. In the same way that binary circuits reject noise by using two voltage levels which are spaced further apart than the uncertainty due to noise, digital transmission combats time uncertainty by using events, known as transitions, spaced apart at integer multiples of some basic time period, called a detent, which is larger than the typical time uncertainty. Figure 3.8 shows how this jitter-rejection mechanism works.

As ideal transitions occur at multiples of a basic period, an oscilloscope which is repeatedly triggered on a channel-coded signal carrying random data will show an eye pattern if connected to the output of the equalizer. Study of the eye pattern reveals how well the coding used suits the channel. With a short cable, the losses will be small, and the eye opening will be virtually square except for some edge sloping due to cable capacitance. As cable length increases, the harmonics are lost and the remaining fundamental gives the eyes a diamond shape. Noise closes the eyes in a vertical direction, and jitter closes the eyes in a horizontal direction, as in Figure 3.9. If the eyes remain sensibly open, this will be possible. Clearly more jitter can be tolerated if there is less noise, and vice versa. If the equalizer is adjustable, the optimum setting will be where the greatest eye opening is obtained.

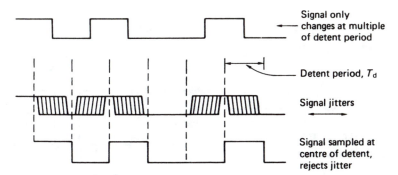

Signal only changes at multiple of detent period

Detent period, T_d

Signal jitters

Signal sampled at centre of detent, rejects jitter

Figure 3.8 A certain amount of jitter can be rejected by changing the signal at multiples of the basic detent period T_d

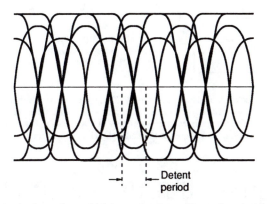

Detent period

Figure 3.9 A transmitted waveform which is generated according to the principle of Figure 3.8 will appear like this on an oscilloscope as successive parts of the waveform are superimposed on the tube. When the waveform is rounded off by losses, diamond-shaped eyes are left in the centre, spaced apart by the detent period

In the centre of the eyes, at regular intervals, the receiver must make binary decisions about the state of the signal, high or low, using the slicer output. The receiver is in fact sampling the output of the slicer, and it needs to have a sampling clock in order to do that. The clock edges which operate the sampler must be in the centre of the eyes.

A fixed frequency clock would be of no use since, even if it was sufficiently stable, it would not know what phase to run at.

The only way in which the sampling clock can be obtained is to use a phase-locked loop to regenerate it from the clock content of the self-clocking channel-coded waveform. In phase-locked loops, the voltage-controlled oscillator is driven by a phase error measured between the output and some reference, such that the oscillator eventually runs at the same frequency as the reference. If a divider is placed between the VCO and the phase comparator, as in Figure 3.10, the VCO frequency can be made to be a multiple of the reference. This also has the effect of making the loop more heavily damped. If a channel-coded waveform is used as a reference to a PLL, the loop will be able to make a phase comparison whenever a transition arrives, but when there are several detents between transitions, the loop

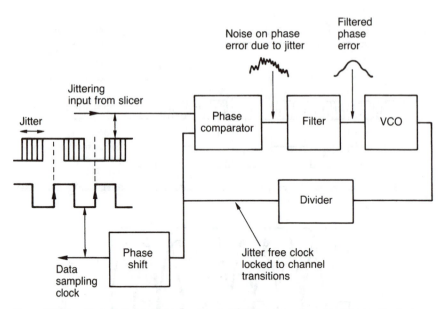

Figure 3.10 In a phase-locked loop data receiver the VCO frequency is synchronized to the clock content of the transmitted waveform so the clock can continue when transitions are omitted to send data. The filter in the phase error removes jitter from the VCO. A suitable phase shift produces sampling edges which are used to measure the incoming signal level between transitions so that jitter is rejected

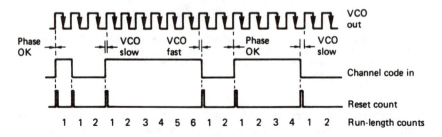

Figure 3.11 In order to reconstruct the channel patterns, a phase-locked loop is fed with the channel code and freewheels between transitions, correcting its phase at each one. Counting the low-going VCO edges between transitions reconstructs the channel bits. If the data rate changes the VCO will track. If the maximum run length is too long, the VCO will not be able to phase-correct often enough and may miscount channel bits in the presence of jitter

will flywheel at the last known frequency and phase until it can rephase at a subsequent transition. Once the loop is locked, clock edges will be phased with the average phase of the jittering edges of the input waveform. If, for example, falling edges of the clock are phased to input transitions, then rising edges will be in the centre of the eyes and can be used to operate the sampling process in such a way that the maximum jitter is rejected. Cycles of the VCO can be counted to measure the number of detents between transitions and hence to decode the information. Figure 3.11 illustrates this mechanism.

Note that certain types of phase detector, such as the phase-frequency detector, are not suitable for flywheeling circuits as they count the number of pulse transitions in a given time period.

Clearly data cannot be separated if the PLL is not locked, but it cannot be locked until it has seen transitions for a reasonable period. In recorders, which have discontinuous recorded blocks to allow editing, the solution is to precede each data block with a pattern of transitions whose sole purpose is to provide a timing reference for synchronizing the phase-locked loop. This pattern is known as a preamble. In interfaces, the transmission can be continuous and there is no difficulty remaining in lock indefinitely. There will simply be a short delay on first applying the signal before the receiver locks to it.

It is important for reliable data reception that the phase-locked loop in a receiver is correctly centred. This means that the free running frequency of the oscillator is very close to the frequency at which it will run when the loop is locked. If the VCO is not centred, a static control voltage will be required to pull it back to the correct frequency. This can only be obtained with a static phase error. This phase error means that the incoming waveform is not being sampled in the centres of the eyes, but with an offset, resulting in the ability to reject jitter being impaired.

There are a number of ways in which the VCO centring adjustment can be made, according to the equipment and the test gear available. If a digital frequency meter is available, it can be used to measure the VCO frequency. This is then adjusted, without input, until the correct value is obtained.

Another approach is to monitor the VCO control voltage and to adjust the centring until there is little or no change to the control voltage when the input is disconnected or reconnected.

Finally, if an error detection system is available it is also possible to centre the VCO by adjusting for minimum error rate on an input signal which has been artificially impaired by the insertion of a cable simulator. Following the adjustment the impairment is then removed.

3.3 Channel coding

3.3.1 Overview

In summary, it is not practicable simply to serialize raw data in a shift register for the purpose of transmission except over relatively short distances. Practical systems require the use of a modulation scheme, known as a channel code, which expresses the data as waveforms which are self-clocking in order to reject jitter, separate the received bits and to avoid skew on separate clock lines. The coded waveforms should further be DC free or nearly so to enable slicing in the presence of losses and have a narrower spectrum than the raw data to make equalization possible.

Jitter causes uncertainty about the time at which a particular event occurred. The frequency response of the channel then places an overall limit on the spacing of events in the channel. Particular emphasis must be placed on the interplay of bandwidth, jitter and noise, which will be shown here to be the key to the design of a successful channel code.

Figure 3.12 shows that a channel coder is necessary prior to the record stage, and that a decoder, known as a data separator, is necessary after the replay stage. A convenient definition of a channel code (for there are certainly others) is: 'A method

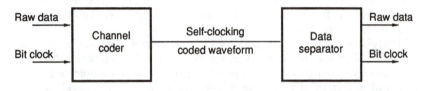

Figure 3.12 Channel coding requires a coder at the transmitter which combines a bit clock with the serial data to produce a channel-coded waveform. The receiver has a data separator which extracts the clock content of the waveform and decodes it to serial data

of modulating real data such that it can be reliably received despite the shortcomings of a real channel, while making maximum economic use of the channel capacity.'

The basic time periods of a channel-coded waveform are called positions or detents, in which the transmitted voltage will be reversed or stay the same. The symbol used for the units of channel time is T_d.

There are many ways of creating such a waveform, but the most convenient is to convert the raw data bits to a larger number of channel bits which are output from a shift register at the detent rate. The coded waveform will then be high or low according to the state of a channel bit which describes the detent.

Channel coding is the art of converting real data into channel bits. It is important to appreciate that the convention most commonly used in coding is one in which a channel-bit 1 represents a voltage change, whereas a 0 represents no change. This convention is used because it is possible to assemble sequential groups of channel bits together without worrying about whether the polarity of the end of the last group matches the beginning of the next. The polarity is unimportant in most codes and all that matters is the length of time between transitions.

One of the fundamental parameters of a channel code is the density ratio (DR). One definition of density ratio is that it is the worst-case ratio of the number of data bits recorded to the number of transitions in the channel. It can also be thought of as the ratio between the Nyquist rate of the data and the frequency response of the channel. When better hardware is available to increase the capacity of a channel, the use of a higher density ratio code multiplies the capacity further.

As jitter is such an important issue in digital recording, a parameter has been introduced to quantify the ability of a channel code to reject time instability. This parameter, the jitter margin, also known as the window margin or phase margin (T_w), is defined as the permitted range of time over which a transition can still be received correctly, divided by the data bit-cell period (T).

Since equalization is often difficult in practice, a code which has a large jitter margin will sometimes be used because it resists the effects of intersymbol interference well. Such a code may achieve a better performance than a code with a higher density ratio but poor jitter performance.

A more realistic comparison of code performance will be obtained by taking into account both density ratio and jitter margin. This is the purpose of the figure of merit (FoM), which is defined as $DR \times T_w$.

3.3.2 FM code

The FM code, also known as Manchester code or bi-phase mark code, shown in Figure 3.13 was the first practical self-clocking binary code and it is suitable for both transmission and recording. It is DC free and very easy to encode and decode. It is the code specified for the AES/EBU digital audio interconnect standard which will be described in Chapter 4. In the field of recording it remains in use today only where density is not of prime importance, for example in SMPTE/EBU timecode for professional audio and video recorders.

In FM there is always a transition at the bit-cell boundary which acts as a clock. For a data 1, there is an additional transition at the bit-cell centre. Figure 3.13 shows that each data bit can be represented by two channel bits. For a data 0, they will be 10, and for a data 1 they will be 11. Since the first bit is always 1, it conveys no information, and is responsible for the density ratio of only one-half. Since there can be two transitions for each data bit, the jitter margin can only be half a bit, and the resulting FoM is only 0.25. The high clock content of FM does, however, mean that data recovery is possible over a wide range of speeds; hence the use for timecode. The lowest frequency in FM is due to a stream of zeros and is equal to half the bit rate. The highest frequency is due to a stream of ones, and is equal to the bit rate. Thus the fundamentals of FM are within a band of one octave. Effective equalization is generally possible over such a band.

Figure 3.13 also shows how an FM coder works. Data words are loaded into the input shift register which is clocked at the data bit rate. Each data bit is converted to two channel bits in the code book or lookup table. These channel bits are loaded into the output register. The output register is clocked twice as fast as the input register because there are twice as many channel bits as data bits. The ratio of the two clocks is called the code rate; in this case it is a rate one-half code. Ones in the serial channel bit output represent transitions whereas zeros represent no change.

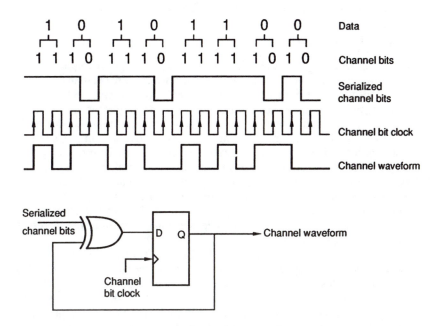

Figure 3.13 In the FM code, every bit commences with a transition, and a second transition will be present to indicate a one. This code is easily generated by creating two channel bits for each data bit. The first one is always one, the second is the same as the data bit. Sending these channel bits to a waveform generator produces the coded waveform

The channel bits are fed to the waveform generator which is a 1 bit delay, clocked at the channel bit rate, and an exclusive OR gate. This changes state when a channel bit 1 is input. The result is a coded FM waveform where there is always a transition at the beginning of the data bit period, and a second optional transition whose presence indicates a one.

3.3.3 4/5 code

In the MADI (Multi-channel Audio Digital Interface) standard, as in FDDI (fibre distributed digital interface), a four-fifths rate code is used where groups of four data bits are represented by groups of five channel bits. Four bits have 16 combinations whereas five bits have 32 combinations. Clearly only 16 out of these 32 are necessary to convey all the possible data. The 16 channel bit patterns chosen are those which have the least DC component combined with a high clock content.

The jitter resistance of a group code is equal to the code rate. for example, in 4/5 transitions cannot be closer than 0.8 of a data bit apart and so this represents the peak-to-peak jitter which can be rejected. The density ratio is also 0.8 so the FoM is 0.64; an improvement over FM.

A further advantage of group coding is that it is possible to have codes which have no data meaning. In MADI further channel bit patterns are used for packing. This is where dummy data is sent when the real data rate is low in order to keep the channel frequencies constant. This is necessary so that fixed equalization can be used. The packing pattern does not decode to data and so it can be easily discarded at the receiver.

Further details of MADI can be found in Chapter 4.

3.3.4 The 8/9 group code

This code was developed to allow serial communication of digital component video sampled in 4:2:2 form. This interface will be discussed in more detail in Chapter 8.

In this code groups of eight data bits are represented by nine channel bits. Unlike previous examples, this code uses a representation where the channel bits represent the state of the waveform directly rather than controlling transitions. There are 512 combinations in nine bits, and those which have either five or four zeros in each symbol are selected in order to prevent excessive DC content. Unfortunately there are only 213 codes with this characteristic, not 256, and therefore 43 codes having a larger DC offset of six bits to three have to be accepted. In practice these unbalanced symbols have a complementary version having the opposite DC content. Figure 3.14 shows the code book where it will be seen that certain codes have two alternatives. The code book is in hexadecimal where one digit represents all values of four bits using 0–9 then A–F.

The coder for 8/9 is shown in Figure 3.15. Eight bits are converted to nine in a lookup table as before, but the lookup table also outputs the code digital sum (CDS) of each code. The CDS is the number of ones in the group minus the number of zeros, which can be ±1 or ±3 in this code. CDS represents the average voltage of the waveform which results when the group is serialized. When a series of groups are serialized, the relative number of ones and zeros up to a given point is known as the digital sum value (DSV) at that point. DSV at the end of a symbol can be obtained by adding together the CDS of all of the previous groups. The coder does this in an accumulator. When a new group is output from the lookup table, the DSV

8B	9B	$\overline{9B}$	8B	9B	$\overline{9B}$	8B	9B	$\overline{9B}$	8B	9B	$\overline{9B}$	8B	9B	$\overline{9B}$	8B	9B	$\overline{9B}$
00	0FE	101	2C	1AC		58	099		84	0D5		B0	11A		DC	131	
01	027		2D	057		59	166		85	12A		B1	0E9		DD	0DC	
02	1D8		2E	09B		5A	09B		86	095		B2	116		DE	127	
03	033		2F	059		5B	164		87	16A		B3	02E		DF	0E2	
04	1CC		30	1A6		5C	09D		88	0B5		B4	1D1		E0	123	
05	037		31	05B		5D	162		89	14A		B5	036		E1	0E4	
06	1C8		32	05D		5E	0A3		8A	09A		B6	1C9		E2	11D	
07	039		33	1A4		5F	15C		8B	165		B7	03A		E3	0E6	
08	1C6		34	065		60	0A7		8C	0A6		B8	1C5		E4	11B	
09	03B		35	19A		61	158		8D	159		B9	04E		E5	0E8	
0A	1C4		36	069		62	025	1DA	8E	0AC		BA	1B1		E6	119	
0B	03D		37	196		63	0A1	15E	8F	153		BB	05C		E7	0EC	
0C	1C2		38	026	1D9	64	029	1D6	90	0AE		BC	1A3		E8	117	
0D	14D		39	08C	173	65	091	16E	91	151		BD	05E		E9	0F2	
0E	0B4		3A	02C	1D3	66	045	1BA	92	02A	1D5	BE	1A1		EA	113	
0F	14B		3B	098	167	67	089	176	93	092	16D	BF	066		EB	0F4	
10	1A2		3C	032	1CD	68	049	1B6	94	04A	1B5	C0	199		EC	10D	
11	0B6		3D	0BE	141	69	085	17A	95	094	16B	C1	06C		ED	076	
12	149		3E	034	1CB	6A	051	1AE	96	0A8	157	C2	193		EE	10B	
13	0BA		3F	0C2	13D	6B	08A	175	97	0B7	148	C3	06E		EF	0C7	
14	145		40	046	1B9	6C	0A4	15B	98	0F5	10A	C4	191		F0	13C	
15	0CA		41	0C4	13B	6D	054	1AB	99	0BB	144	C5	072		F1	047	
16	135		42	04C	1B3	6E	0A2	15D	9A	0ED	112	C6	18D		F2	1B8	
17	0D2		43	0C8	137	6F	052	1AD	9B	0BD	142	C7	074		F3	067	
18	12D		44	058	1A7	70	056		9C	0EB	114	C8	18B		F4	19C	
19	0D4		45	0B1		71	1A9		9D	0D7	129	C9	07A		F5	071	
1A	129		46	14E		72	05A		9E	0DD	122	CA	189		F6	198	
1B	0D6		47	0B3		73	1A5		9F	0DB	124	CB	08E		F7	073	
1C	125		48	14C		74	06A		A0	146		CC	185		F8	18E	
1D	0DA		49	0B9		75	195		A1	0C5		CD	09C		F9	079	
1E	115		4A	06B		76	096		A2	13A		CE	171		FA	18C	
1F	0EA		4B	194		77	169		A3	0C9		CF	09E		FB	087	
20	0B2		4C	06D		78	0A9		A4	136		D0	163		FC	186	
21	02B		4D	192		79	156		A5	0CB		D1	0B8		FD	0C3	
22	1D4		4E	075		7A	0AB		A6	134		D2	161		FE	178	
23	02D		4F	18A		7B	154		A7	0CD		D3	0BC		FF	062	190
24	1D2		50	08B		7C	0A5		A8	132		D4	147				
25	035		51	174		7D	15A		A9	0D1		D5	0C6				
26	1CA		52	08D		7E	0AD		AA	12E		D6	143				
27	04B		53	172		7F	152		AB	0D3		D7	0CC				
28	1B4		54	093		80	155		AC	12C		D8	139				
29	04D		55	16C		81	0AA		AD	0D9		D9	0CE				
2A	1B2		56	097		82	055		AE	126		DA	133				
2B	053		57	168		83	1AA		AF	0E5		DB	0D8				

Figure 3.14 Codebook for CCIR 8/9 code. Some 9 bit symbols have excessive DC content, and a complemented version $\overline{9B}$ also exists. Normal and complemented symbols are used alternately to control DC

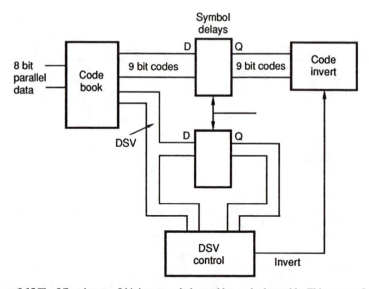

Figure 3.15 The 8/9 coder uses 8 bit input symbols to address a lookup table. This outputs 9 bit channel symbols as well as their DC content or code digital sum, CDS. The digital sum value (DSV) controller selects the polarity of the 9 bit code in order to maintain the waveform as nearly DC free as possible

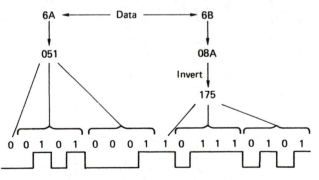

Figure 3.16 In this example, the symbol on the left results in a code symbol having six zeros, which has negative DC content. The next symbol which has a DC content will be inverted as shown so that the DC content is cancelled

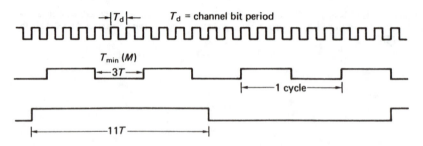

Figure 3.17 A channel code's spectrum can be controlled by placing limits on T_{min} (M) and T_{max} which define upper and lower frequencies. Ratio of T_{max} to T_{min} determines asymmetry of waveform and predicts DC content and peak shift

accumulator compares the current DSV with what the next DSV would be. If adding the new CDS improves the DSV, no action is taken, but if it is made worse, the new code is inverted. The DC component of the new code then has the opposite polarity.

Figure 3.16 illustrates the operation where the second code is inverted so that its DC content balances that of the first. It will be seen that the resultant waveform has nine ones and nine zeros and is therefore perfectly DC free overall even if individual groups deviate. Such a code is called a low disparity code to distinguish it from an unconditionally DC-free code.

Figure 3.17 shows that the lower spectral limit is influenced by the maximum distance between transitions T_{max}. This is also obtained by multiplying the maximum number of detent periods between transitions by the code rate.

The length of time between channel transitions is known as the run length. Another name for this class is the run-length-limited (RLL) codes[1]. It is, however, possible for a code to have run-length limits without it being a group code.

3.4 Error detection

Error detection works by adding some bits to the data which are calculated from the data. This creates an entity called a codeword which spans a greater length of time than one bit alone. The statistics of noise means that whilst one bit may be lost in a codeword, the loss of the rest of the codeword because of noise is highly improbable. The greater the timespan over which the coding is performed, the greater will be the reliability achieved, and systems are then able to handle burst errors as well as those due to noise. This does mean, however, that delay will be experienced on reception.

Figure 3.18 shows the basics of all error detection systems. At the transmitter, data to be sent is also passed to a coder which performs some calculation on the data. The result of the calculation is added to the end of the data. The receiver will perform the same calculation and compare the locally calculated result with the transmitted result. Should the two be different, an error has occurred.

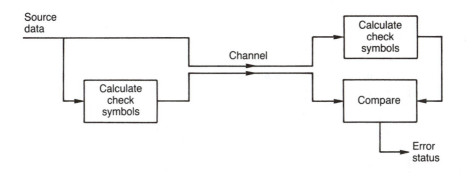

Figure 3.18 All error detection depends upon some additional symbols which are calculated from and appended to the data at the transmitter. The receiver makes a similar calculation and compares the results

3.4.1 Parity

The exclusive OR gate shows up extensively in error detection and coding. One way of remembering the characteristics of this useful device is that there will be an output when the inputs are different. Inspection of the truth table in Figure 3.19 will show that there is an even number of ones in each row (zero is an even number) and so the device could also be called an even parity gate.

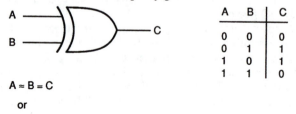

A	B	C
0	0	0
0	1	1
1	0	1
1	1	0

$A \approx B = C$

or

$(A + B) \bmod 2 = C$

Figure 3.19 The exclusive OR gate and its truth table. The action of the gate is easy to remember as the output is true when the inputs are different

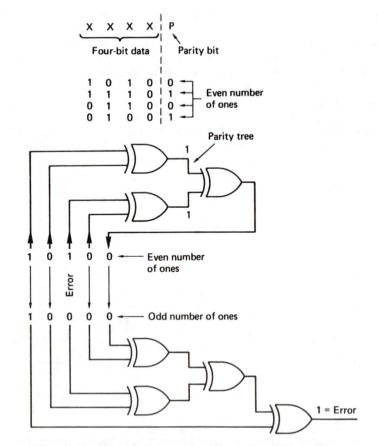

Figure 3.20 Parity checking adds up the number of ones in a word using, in this example, parity trees. One error bit and odd numbers of errors are detected. Even numbers of errors cannot be detected

Parity is a fundamental concept in error detection. In Figure 3.20, the example is given of a 4 bit data word which is to be protected. If an extra bit is added to the word which is calculated in such a way that the total number of ones in the 5 bit word is even, this property can be tested on receipt. The generation of the parity bit in Figure 3.20 can be performed by a number of the ubiquitous XOR gates configured into what is known as a parity tree. In the figure, if a bit is corrupted, the received message will be seen no longer to have an even number of ones. If two bits are corrupted, the failure will be undetected. This example can be used to introduce much of the terminology of error detection. The extra bit added to the message carries no information of its own, since it is calculated from the other bits. It is therefore called a redundant bit. The addition of the redundant bit gives the message a special property, i.e. the number of ones is even. A message having some special property irrespective of the actual data content is called a codeword. All error correction relies on adding redundancy to real data to form codewords for transmission. If any corruption occurs, the intention is that the received message will not have the special property; in other words if the received message is not a codeword there has definitely been an error. If the received message is a codeword, there probably has not been an error. The word 'probably' must be used because the figure shows that two bits in error will cause the received message to be a codeword, which cannot be discerned from an error-free message. If it is known that generally the only failure mechanism in the channel in question is loss of a single bit, it is *assumed* that receipt of a codeword means that there has been no error. If there is a probability of two error bits, that becomes very nearly the probability of failing to detect an error, since all odd numbers of errors will be detected, and a four-bit error is much less likely. It is paramount in all error-detection systems that the protection used should be appropriate for the probability of errors to be encountered. Another result demonstrated by the example is that we can only guarantee to detect the same number of bits in error as there are redundant bits.

In the example of Figure 3.20, the error was detected but it was not possible to say which bit was in error. This is perfectly adequate for a digital interface using cables where the signal quality of a given installation tends to be stable and errors are highly infrequent. An example of this is the AES/EBU digital audio interface where the error detection in the audio sample data uses simple parity. A parity error is an indication that the system is faulty or marginal in operation; normally no errors are experienced and error correction is unnecessary, especially as it causes delay.

3.4.2 Hamming code

A single parity check cannot locate the position of an error, but this can be determined by using more parity checks. In Figure 3.21, the four data bits have been used to compute three redundancy bits, making a 7 bit codeword. The four data bits are examined in turn, and each bit which is a 1 will cause the corresponding row of a generator matrix to be added to an exclusive OR sum. For example, if the data were 1001, the top and bottom rows of the matrix would be XORed. The matrix used is known as an identity matrix, because the data bits in the codeword are identical to the data bits to be conveyed. This is useful because the original data can be stored unmodified, and the check bits are simply attached to the end to make a so-called systematic codeword. Almost all digital recording equipment uses systematic codes. The way in which the redundancy bits are calculated is simply that they do not all use every data bit. If a data bit has not been included in a parity check, it can

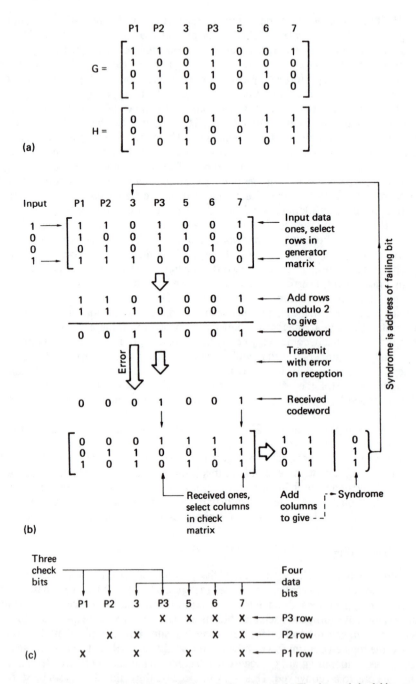

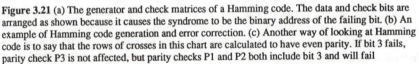

Figure 3.21 (a) The generator and check matrices of a Hamming code. The data and check bits are arranged as shown because it causes the syndrome to be the binary address of the failing bit. (b) An example of Hamming code generation and error correction. (c) Another way of looking at Hamming code is to say that the rows of crosses in this chart are calculated to have even parity. If bit 3 fails, parity check P3 is not affected, but parity checks P1 and P2 both include bit 3 and will fail

fail without affecting the outcome of that check. The position of the error is deduced from the pattern of successful and unsuccessful checks in the check matrix. This pattern is known as a syndrome. Since the code is binary, only the location of the error needs to be found. Correction of the failed bit is made simply by inversion.

In the figure the example of a failing bit is given. Bit three fails, and because this bit is included in only two of the checks, there are two ones in the failure pattern, 011. As some care was taken in designing the matrix pattern for the generation of the check bits, the syndrome, 011, is the address of the failing bit. This is the fundamental feature of the Hamming codes due to Richard Hamming[2]. The performance of this 7 bit codeword can be assessed. In seven bits there can be 128 combinations, but in four data bits there are only sixteen combinations. Thus out of 128 possible received messages, only sixteen will be codewords, so if the message is completely trashed by a gross corruption, it will still be possible to detect that this has happened 112 times out of 127, as in these cases the syndrome will be non-zero (the 128th case is the correct data). There is thus only a probability of detecting that all of the message is corrupt. In an idle moment it is possible to work out, in a similar way, the number of false codewords which can result from different numbers of bits being assumed to have failed. For less than three bits, the failure will always be detected, because there are three check bits.

The Hamming code is used in the component digital video interface to protect vital synchronizing words. The code words are very short to minimize processing delay. More details of the application will be found in Chapter 8

3.4.3 Cyclic codes

The implementation of a Hamming code can be made very fast using parity trees or ROM, which is ideal for the digital video applications mentioned above. However, in most digital transmission applications, the data is transmitted serially and relatively large data blocks are used. Where large data blocks are to be handled, the use of a lookup table or tree has to be abandoned because it would become impossibly large. The principle of generator and check matrices will still be employed, but they will be matrices which can be generated by an algorithm.

Where data can be accessed serially, simpler circuitry can be used because the same gate will be used for many XOR operations. Unfortunately the reduction in component count is offset by an increase in the complexity of the process.

The circuit of Figure 3.22 is a kind of shift register, but with a peculiar feedback arrangement which leads it to be known as a twisted-ring counter. If seven message bits A–G are applied serially to this circuit, and each one of them is clocked, the outcome can be followed in the diagram. As bit A is presented and the system is clocked, bit A will enter the left-hand latch. When bits B and C are presented, A moves across to the right. Both XOR gates will have A on the upper input from the right-hand latch: the left one has D on the lower input and the right one has B on the lower input. When clocked, the left latch will thus be loaded with the XOR of A and D, and the right one with the XOR of A and B. The remainder of the sequence can be followed, bearing in mind that when the same term appears on both inputs of an XOR gate, it goes out, as the exclusive OR of something with itself is nothing. At the end of the process, the latches contain three different expressions. Essentially, the circuit makes three parity checks through the message, leaving the result of each in the three stages of the register. In the figure, these expressions have been used to draw up a check matrix. The significance of these steps can now be explained. The

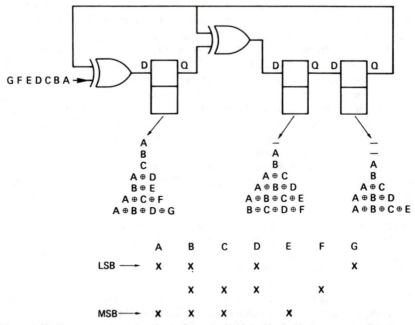

Figure 3.22 When seven successive bits A–G are clocked into this circuit, the contents of the three latches are shown for each clock. The final result is a parity-check matrix

bits A B C and D are four data bits, and the bits E F and G are redundancy. When the redundancy is calculated, bit E is chosen so that there is an even number of ones in bits A B C and E; bit F is chosen such that the same applies to bits B C D and F, and similarly for bit G. Thus the four data bits and the three check bits form a 7 bit codeword. If there is no error in the codeword, when it is fed into the circuit shown, the result of each of the three parity checks will be zero and every stage of the shift register will be cleared. If a bit in the codeword is corrupted, there will be a non-zero result. For example, if bit D fails, the check on bits A B D and G will fail, and a 1 will appear in the left-hand latch. The check on bits B C D F will also fail, and the centre latch will set. The check on bits A B C E will not fail, because D is not involved in it, making the right-hand bit zero. There will be a syndrome of 110 in the register, and this will be seen from the check matrix to correspond to an error in bit D. Whichever bit fails, there will be a different 3 bit syndrome which uniquely identifies the failed bit. As there are only three latches, there can be eight different syndromes. One of these is zero, which is the error-free condition, and so there are seven remaining error syndromes. The length of the codeword cannot exceed seven bits, or there would not be enough syndromes to correct all of the bits. This can also be made to tie in with the generation of the check matrix. If fourteen bits, A to N, were fed into the circuit shown, the result would be that the check matrix repeated twice, and if a syndrome of 101 were to result, it could not be determined whether bit D or bit K failed. Because the check repeats every seven bits, the code is said to be a cyclic redundancy check (CRC) code.

In Figure 3.21 an example of a Hamming code was given. Comparison of the check matrix of Figure 3.22 with that of Figure 3.21 will show that the only difference is the order of the matrix columns. The two different processes have thus achieved exactly the same results, and the performance of both must be identical. In practice Hamming code blocks will generally be much smaller than the blocks used in CRC codes.

It has been seen that the circuit shown makes a matrix check on a received word to determine if there has been an error, but the same circuit can also be used to generate the check bits. To visualize how this is done, examine what happens if only the data bits A B C and D are known, and the check bits E F and G are set to zero. If this message, ABCD000, is fed into the circuit, the left-hand latch will afterwards contain the XOR of A B C and zero, which is of course what E should be. The centre latch will contain the XOR of B C D and zero, which is what F should be and so on. This process is not quite ideal, however, because it is necessary to wait for three clock periods after entering the data before the check bits are available. Where the data is simultaneously being recorded and fed into the encoder, the delay would prevent the check bits being easily added to the end of the data stream. This problem can be overcome by slightly modifying the encoder circuit as shown in Figure 3.23. By moving the position of the input to the right, the operation of the circuit is advanced so that the check bits are ready after only four clocks. The process can be followed in the diagram for the four data bits A B C and D. On the first clock, bit A enters the left two latches, whereas on the second clock, bit B will appear on the upper input of the left XOR gate, with bit A on the lower input, causing the centre latch to load the XOR of A and B and so on.

The way in which the correction system works has been described in engineering terms, but it can be described mathematically if analysis is contemplated. The mathematical principle is understood by recognizing the shift register described above as a divider.

Just as the position of a decimal digit in a number determines the power of ten (whether that digit means one, ten or a hundred), the position of a binary digit determines the power of two (whether it means one, two or four). It is possible to rewrite a binary number so that it is expressed as a list of powers of two. For example, the binary number 1101 means $8 + 4 + 1$, and can be written:

$$2^3 + 2^2 + 2^0$$

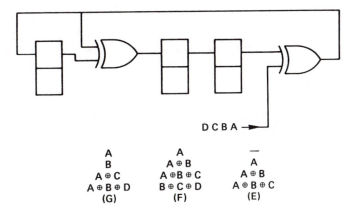

Figure 3.23 By moving the insertion point three places to the right, the calculation of the check bits is completed in only four clock periods and they can follow the data immediately. This is equivalent to multiplying the data by x^3

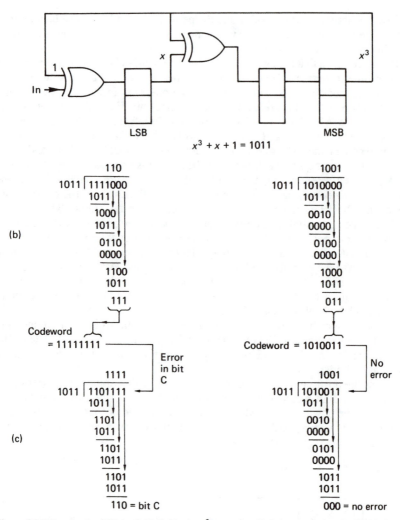

Figure 3.24 The circuit of Figure 3.22 divides by $x^3 + x + 1$ to find the remainder. At (b) this is used to calculate the check bits. At (c) left, there is an error, and the non-zero syndrome 110 points to bit C. At (c) right, the syndrome is zero, and no error is indicated

In fact, much of the theory of error correction applies to symbols in number bases other than 2, so that the number can also be written more generally as:

$$x^3 + x^2 + 1 \quad (2^0 = 1)$$

which also looks much more impressive. This expression, containing as it does various powers, is of course a polynomial, and the circuit of Figure 3.22 which has been seen to construct a parity-check matrix on a codeword can also be described as calculating the remainder due to dividing the input by a polynomial using modulo-2 arithmetic. In modulo-2 there are no borrows or carries, and addition and subtraction are replaced by the XOR function, which makes hardware implementation very easy. In Figure 3.24 it will be seen that the circuit of Figure 3.22 actually divides the codeword by a polynomial which is:

$$x^3 + x + 1 \quad \text{(or 1011)}$$

This can be deduced from the fact that the right-hand bit is fed into several lower-order stages of the register at once. Once all the bits of the message have been clocked in, the circuit contains the remainder. In mathematical terms, the special property of a codeword is that it is a polynomial which yields a remainder of zero when divided by the generating polynomial. If an error has occurred it is considered that this is due to an error polynomial which has been added to the codeword polynomial. If a codeword divided by the check polynomial is zero, a non-zero syndrome must represent the error polynomial divided by the check polynomial. Some examples of modulo-2 division are given in Figure 3.24 which can be compared with the parallel computation of parity checks according to the matrix of Figure 3.22.

If a codeword has to give zero remainder when divided, it follows that the data can be converted to a codeword by adding the remainder when the data are divided. Generally speaking the remainder would have to be subtracted, but in modulo-2 there is no distinction. This process is also illustrated in Figure 3.24. The four data bits have three zeros placed on the right-hand end, to make the wordlength equal to that of a codeword, and this word is then divided by the polynomial to calculate the remainder. The remainder is added to the zero-extended data to form a codeword. The modified circuit of Figure 3.23 can be described as premultiplying the data by x^3 before dividing.

Cyclic codes are of primary importance for detecting errors, and several have been standardized for use in digital communications. The most common of these are:

$$x^{16} + x^{15} + x^2 + 1 \quad \text{(CRC-16)}$$

$$x^{16} + x^{12} + x^5 + 1 \quad \text{(CRC-CCITT)}$$

The code:

$$x^8 + x^4 + x^3 + x^2 + x^0$$

is also frequently used.

The EDH (error detection and handling) option of the SDI (serial digital interface) uses the CRC-CCITT code and further details of this application will be found in Chapter 8. Cyclic codes will also be used in the technique of signature analysis which is used for data reliability testing of digital interfaces. Signature analysis will also be treated in Chapter 8.

The AES/EBU digital audio interface described in Chapter 4 uses an 8 bit cyclic code to protect the channel-status data. This error detection is needed because the AES/EBU interface may be used with a routing switcher in a large system. The switcher may change configuration from time to time and select a different source for a given output. If the inputs are all synchronous, the switch can be made between audio samples and no audio data is corrupted. However, the channel status message is sent very slowly and takes several milliseconds to complete a block. A changing router configuration will corrupt the channel status block and this must be detected. Correction is unnecessary as it is possible to wait for the next error-free block. The polynomial used and a typical circuit for generating it can be seen in Figure 3.25.

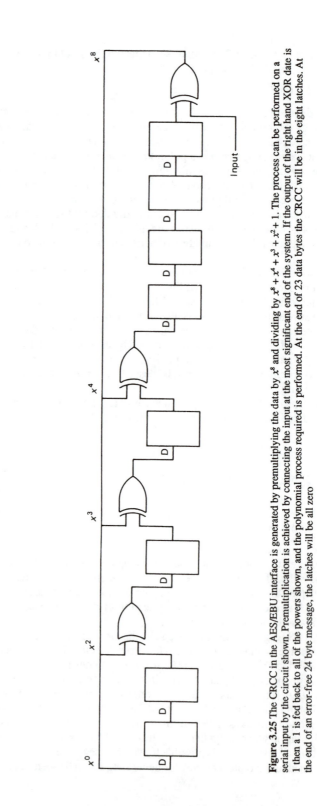

Figure 3.25 The CRCC in the AES/EBU interface is generated by premultiplying the data by x^8 and dividing by $x^8 + x^4 + x^3 + x^2 + 1$. The process can be performed on a serial input by the circuit shown. Premultiplication is achieved by connecting the input at the most significant end of the system. If the output of the right hand XOR date is 1 then a 1 is fed back to all of the powers shown, and the polynomial process required is performed. At the end of 23 data bytes the CRCC will be in the eight latches. At the end of an error-free 24 byte message, the latches will be all zero

3.4.4 The Galois field

It is interesting to study the operation of the circuit of Figure 3.24 with no input. Whatever the starting condition of the three bits in the latches, the same state will always be reached again after seven clocks, except if zero is used. The states of the latches form an endless ring of non-sequential numbers called a Galois field after the French mathematical prodigy Evariste Galois who discovered them. With the correct choice of generator polynomial, the states of the circuit can form a maximum length sequence so there are as many states as are permitted by the wordlength, hence the term cyclic code. As the states of the sequence have many of the characteristics of random numbers, yet are repeatable, the result can also be called a pseudo-random sequence. As the all-zeros case is disallowed, the length of a maximum length sequence generated by a register of m bits cannot exceed (2^m-1) states. This figure will also be the length of the codeword in optimum cyclic codes. The Galois field, however, includes the zero term. It is useful to explore the bizarre mathematics of Galois fields which use modulo-2 arithmetic. Familiarity with such manipulations is helpful when studying the randomizing used in NICAM and the scrambled serial interface, SDI.

As the circuit of Figure 3.24 divides the input by the function $F(x) = x^3 + x + 1$ and there is no input in this case, the operation of the circuit has to be described by:

$$x^3 + x + 1 = 0$$

Each 3 bit state of the circuit can be described by combinations of powers of x, such as:

$$x^2 = 100$$

$$x = 010$$

$$x^2 + x = 110, \text{ etc.}$$

To avoid confusion, the 3 bit state of the field will be called a, which is a primitive element. Now:

$$a^3 + a + 1 = 0$$

In modulo 2:

$$a + a = a^2 + a^2 = 0$$

$$a = x = 010$$

$$a^2 = x^2 = 100$$

$$a^3 = a + 1 = 011$$

$$a^4 = a \times a^3 = a(a + 1) = a^2 + a = 110$$

$$a^5 = a^2 + a + 1 = 111$$

$$a^6 = a \times a^5 = a(a^2 + a + 1)$$

$$= a^3 + a^2 + a = a + 1 + a^2 + a$$

$$= a^2 + 1 = 101$$

$$a^7 = a(a^2 + 1) = a^3 + a \quad = a + 1 + a = 1 = 001$$

In this way it can be seen that the complete set of elements of the Galois field can be expressed by successive powers of the primitive element. Note that the twisted-ring circuit of Figure 3.23 simply raises a to higher and higher powers as it is clocked; thus the seemingly complex multibit changes caused by a single clock of the register become simple to calculate using the correct primitive and the appropriate power.

3.5 Randomizing in NICAM

Figure 3.26 shows how a twisted-ring counter is used to generate the randomizing sequence used in NICAM. A 9 bit device has a sequence length of 2^9-1, and is preset to 111111111 at the beginning of each frame. The polynomial is $x^9 + x^4 + 1$. The serialized data is XORed with the LSB of the Galois field, which randomizes the output which then goes to the modulator. On reception, the derandomizer must contain the identical ring counter which must be reset to the starting condition to bit accuracy. Its output is then added to the data stream from the demodulator. The randomizing will effectively then have been added twice to the data in modulo 2, and as a result is cancelled out leaving the original serial data.

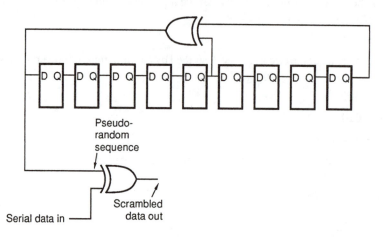

Figure 3.26 In NICAM, randomizing is done by adding the serial data to a polynomial generated by the circuit shown here. The receiver needs an identical system which is synchronous with that of the transmitter

3.6 Convolutional randomizing

The randomizing in NICAM is block based, since this matches the 1 millisecond block structure of the transmission. Where there is no obvious block structure, convolutional or endless randomizing can be used. This is the approach used in the SDI (serial digital interface) described in Chapter 8, which allows composite or component video of up to 10 bit wordlength to be sent serially.

In convolutional randomizing, the signal sent down the channel is the serial data waveform which has been convolved with the impulse response of a digital filter. On reception the signal is deconvolved to restore the original data. Figure 3.27(a)

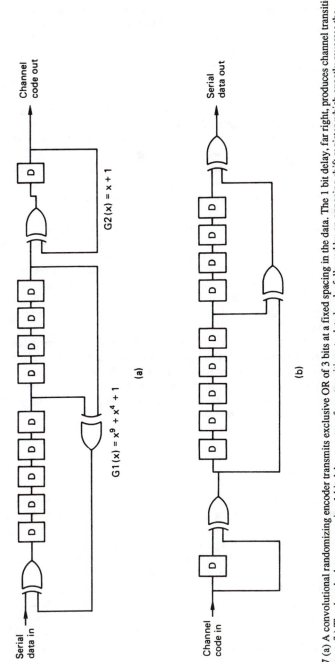

Figure 3.27 (a) A convolutional randomizing encoder transmits exclusive OR of 3 bits at a fixed spacing in the data. The 1 bit delay, far right, produces channel transitions from data ones. (b) The decoder has an opposing 1 bit delay to return from transitions to data levels, followed by an opposing shift register which exactly reverses the coding process

shows that the filter is an infinite impulse response (IIR) filter which has recursive paths from the output back to the input. As it is a 1 bit filter its output cannot decay, and once excited, it runs indefinitely. The filter is followed by a transition generator which consists of a 1 bit delay and an exclusive OR gate. An input 1 results in an output transition on the next clock edge. An input 0 results in no transition.

A result of the infinite impulse response of the filter is that frequent transitions are generated in the channel which result in sufficient clock content for the phase-locked loop in the receiver.

Transitions are converted back to ones by a differentiator in the receiver. This consists of a 1 bit delay with an exclusive OR gate comparing the input and the output. When a transition passes through the delay, the input and the output will be different and the gate outputs a 1 which enters the deconvolution circuit.

Figure 3.27(b) shows that in the deconvolution circuit a data bit is simply the exclusive OR of a number of channel bits at a fixed spacing. The deconvolution is implemented with a shift register having the exclusive OR gates connected in a reverse pattern to that in the encoder. The same effect as block randomizing is obtained, in that long runs are broken up and the DC content is reduced, but it has the advantage over block randomizing that no synchronizing is required to remove the randomizing, although it will still be necessary for deserialization.

It is a characteristic of the convolutional code described that if a bit in the channel (i.e. the cable) is corrupted, then it will cause more than one bit of data to be incorrect. This is known as error propagation. In the case of SDI, up to five data bits can be corrupted by a single channel error and so there is a strong possibility that errors will be visible.

Convolutional codes are also prone to data bit sequences which happen to replicate the feedback sequence at the input gate of the encoder. This results in a long run-length on the cable without transitions and makes it difficult for the receiving PLO to stay in lock, hence the term pathological sequence. Whilst pathological sequences are relatively rare in real program material, they can be generated deliberately by reliability test equipment in order to stress equipment beyond its service conditions.

3.7 Synchronizing

In a serial data receiver, once the PLL has locked to the clock content of the transmission, a serial channel bit stream and a channel bit clock will emerge from the sampler. In a group code, it is essential to know where a group of channel bits begins in order to assemble groups for decoding to data bit groups. In a block randomizing system it is equally vital to know at what point in the serial data stream the words or samples commence. In serial transmission, channel bit groups or randomized data words are sent one after the other, one bit at a time, with no spaces in between, so that although the designer knows that a data block contains, say, 128 bytes, the receiver simply finds 1024 bits in a row. If the exact position of the first bit is not known, then it is not possible to put all the bits in the right places in the right bytes. The effect of sync slippage is devastating, because a 1 bit disparity between the bit count and the bit stream will corrupt every symbol in the block.

The synchronization of the data separator and the synchronization to the block format are two distinct problems, which are often solved by the same sync pattern. The sync pattern is either identical for every block or has a restricted number of versions and it will be recognized by the replay circuitry and used to reset the bit count through the block.

By counting channel bits and dividing by the group size, groups can be separated and decoded to data groups. In a randomized system, counting bits from the sync pattern and dividing by the wordlength enables the replay circuitry to determine the position of the boundaries between data words.

3.7.1 Sync in digital video

Clearly the sync pattern must be one that cannot occur in encoded real data or false synchronizing could occur. It is not sufficient to ensure that no word in the data has the sync pattern; it is necessary to ensure that the sync pattern cannot be created at the junction of two adjacent data words by concatenation.

In digital video such a step is quite easy because the quantizing range of video is restricted such that a sample value of all ones or all zeros cannot occur in the active line. These values correspond to the extreme ends of the pure binary quantizing range and can be reserved for synchronizing without difficulty. This is why the bipolar colour difference signals in component video use offset binary coding instead of two's complement. In two's complement, blanking level would produce a code of all zeros.

The sync pattern of the scrambled serial interface consists of ten data ones followed by twenty data zeros (component) or thirty data zeros (composite). Such a pattern cannot occur due to any sample value or to the concatenation of any valid pair of samples.

3.7.2 Sync in digital audio

In digital audio the two's complement coding scheme is universal and traditionally no codes have been reserved for synchronizing; they are all available for sample values. It would in any case be impossible to reserve all ones or all zeros as these are in the centre of the range in two's complement. It is thus not practical to use a sample code for synchronizing.

In run-length-limited codes this is not a problem. The sync pattern is no longer a data bit pattern but is a specific waveform. If the sync waveform contains run lengths which violate the normal coding limits, there is no way that these run lengths can occur in encoded data, nor any possibility that they will be interpreted as data. They can, however, be readily detected by the replay circuitry. The sync patterns of the AES/EBU interface are shown in Figure 4.3. It will be seen from Figure 3.13 that the maximum run length in FM coded data is 1 bit. The sync pattern begins with a run length of 1.5 bits which is unique. There are three types of sync pattern in the AES/EBU interface, as will be seen in Chapter 4. These are distinguished by the position of a second pulse after the run length violation.

3.7.3 Sync in group codes

In a group code there are many more combinations of channel bits than there are combinations of data bits. Thus after all data bit patterns have been allocated group patterns, there are still many unused group patterns which cannot occur in the data. With care, group patterns can be found which cannot occur due to the concatenation of any pair of groups representing data. These are then unique and can be used for synchronizing.

In MADI, this approach is used as will be seen in Chapter 4.

In the 8/9 serial digital video code, the technique is unnecessary as video has two data codes reserved for synchronizing. These result in special groups which are detected for synchronizing purposes. More details of this process can be found in Chapter 7.

For a more detailed treatment of coding principles the reader is referred elsewhere[3,4]

References

1   TANG, D. T. (1969) Run-length-limited codes. *IEEE International Symposium on Information Theory*

2   HAMMING, R. W. (1950) Error-detecting and error-correcting codes. *Bell Syst. Tech. J.*, vol. 26, pp. 147–160

3   SCHOUHAMER IMMINK, K. A. (1991) *Coding Techniques for Digital Recorders*, Prentice Hall, Englewood Cliffs, NJ

4   WATKINSON, J. R. (1994) *The Art of Data Recording*, Focal Press, Oxford

Chapter 4

Digital audio interfaces – I

For the purposes of this book, digital audio interfaces will be divided into two types: those which have been internationally standardized and those which are associated principally with one manufacturer. Some interfaces of the latter type have become widely used by other manufacturers, either because no alternative existed at the time or in order to provide compatibility between devices, but it is important to distinguish *de facto* standards which have arisen due to commercial predominance from the standards formulated by independent international bodies. A further subdivision is also useful, and that is between interfaces carrying one or two channels of audio data and those carrying a large number of channels.

Naturally, since standards are published and widely available, these chapters should be viewed as commentaries upon or illuminations of the published documents, together with guidelines on their implementation and discussions of the problems involved when attempting to interconnect devices digitally. Although some details of standards will be given here, readers are encouraged to read the information contained herein in conjunction with the standards documents themselves (see the references at the end of each chapter), and to note any additions or modifications to the standards which may have arisen since this book was written. As this book is designed to aid understanding of standards in real situations it is not worded like a standards document and therefore does not use the dogmatic language of such documents. In this chapter will be covered the technical details of internationally standardized audio interfaces, and in the next chapter there will be coverage of *de facto* standards and manufacturers' own interfaces. Further chapters follow on practical applications and synchronization.

4.1 Background to audio interface standardization

Within the audio field it is the Audio Engineering Society (AES) that has been the lead body in determining digital interconnect standards. Although the AES is a professional society, and not a standards body as such, its recommendations first published in the AES3-1985 document[1] have formed the basis for all the international standards documents concerning a two-channel digital audio interface. The society has been instrumental in coordinating professional equipment manufacturers' views on interface standards although it has tended to ignore consumer applications to some extent, and this is perhaps one of the principal roots of confusion in the field. The consumer interface was initially developed by Sony and

Philips, alongside the work being done by the AES, and consequently there are many things in common between the professional and consumer implementations. Before setting out to describe the international standard two-channel interface it is important to give a summary of the history of the standard, since it will then be realized how difficult it is to call this interface by one definitive title!

Amongst other organizations which based standards on AES3 recommendations were the American National Standards Institute (ANSI), the European Broadcasting Union (EBU), the International Radio Consultative Committee (CCIR), the International Electrotechnical Commission (IEC), the Electronic Industries Association of Japan (EIAJ) and the British Standards Institute (BSI). Each of these organizations formulated a document describing a standard for a two-channel digital audio interface, and although these documents are all very similar there are often also either subtle or (in some cases) not-so-subtle differences between them. A useful overview of the evolutionary process which resulted in each of these standards may be found in Finger[2]. The documents concerned are as follows: AES3-1985[1], ANSI S4.40-1985, EBU Tech. 3250-E[3] (1985), CCIR Rec. 647 (1986)[4], CCIR Rec. 647 (1990)[5], IEC 958 (1989)[6] (with subsequent annexes), EIAJ CP-340 (1987)[7], EIAJ CP-1201 (1992)[8], and BS 7239 (1989)[9].

As mentioned above, the roots of the consumer format interface were in a digital interface implemented by Sony and Philips for the CD system in 1984. This interface was modelled on the data format of AES3, but used different electrical characteristics (see section 4.2.4), and is often called the SPDIF (Sony–Philips Digital Interface). Although audio data was in the same format as AES3, there were significant differences in the format of non-audio data. In 1987 the EIAJ CP-340 standard subsequently combined professional (Type I) and consumer (Type II) versions of the interface within one document, and included specifications for non-audio data which aimed to ensure compatibility between different consumer devices such as DAT and CD players, but this interface was by no means identical to the original SPDIF, and this has led to some differences between early CD players and later digital devices conforming to CP-340. CP-340 has recently been renumbered, and is now called CP-1201. At the time of writing this standard is not available in English.

Slightly later than CP-340, and continuing the trend to extend the interface standard also to cover consumer applications, the IEC produced a document which eventually appeared in 1989 as IEC 958 – probably the most influential of the standards documents to date. The consumer version (with its subsequent annexes) is an extension of the SPDIF to allow for wider applications than just the CD, and the professional version is essentially the same as AES3. It also allows for an optical connection. As will be seen later, interpreting this standard to the letter seemingly allows the manufacturer to combine either consumer or professional *data* formats with either 'consumer' or 'professional' *electrical* interfaces, because it does not state that a particular electrical interface must be used in conjunction with a particular data format. Considerable confusion thus exists in the industry over whether consumer devices may be interconnected with professional and vice versa, and the answer to this problem is by no means straightforward, as will be discussed.

Concerning key similarities and differences between the other documents listed above, one should note that EBU Tech. 3250-E is a professional standard, the only effective difference from AES3 being the insistence on the use of transformer coupling. Tech. 3250-E was revised in 1991 to define more aspects of the channel status bits, to allow a speech quality coordination channel in the auxiliary bits, to give an improved electrical specification and to specify which aspects of the interface should be implemented in standard broadcast equipment. CCIR Rec. 647 is a professional standard which does not insist on transformers, but is otherwise similar to the EBU standard. The 1990 revision contains some further definitions of certain non-audio bits, including a use of the auxiliary bits for a low quality coordination channel (see section 4.4). BS 7239 is identical to IEC 958. ANSI S4.40 is identical to AES3.

It may reasonably be concluded from the foregoing discussion that a device which claims conformity to AES3, ANSI S4.40, EBU 3250 or CCIR 647 will be a professional device, but that one which claims conformity to IEC 958, EIAJ CP-340, 1201 or BS 7239 could be either consumer or professional. It is conventional in the latter case to specify Type I or II, for professional or consumer, although strictly this does not solve the problem with IEC 958's ambivalence over which electrical interface is 'professional'. Because it will be necessary to refer to differences between these standards in the following text, the cumbersome but necessary term 'standard two-channel interface' will be used wherever generalization is appropriate, rather than the more usual 'AES/EBU'. It should be noted that within Europe there is a tendency for people to refer to the professional interface as the 'EBU/AES interface'.

Owing to the efforts of four UK companies, a further standard was devised to accommodate up to 56 audio channels. Originally called MADI (Multichannel Audio Digital Interface), it is based on the AES3 data format and has now been standardized as AES10-1991[10]. It also appears as an American Standard: ANSI S4.43-1991. This is only a professional interface and does not suffer from the confusions which surround the two-channel interface as outlined above. It is covered further in section 4.8.

4.2 Standard two-channel interface – overview

Common to all the above standards for a two-channel interface is the data format of the subframe containing samples of audio data for each channel. There are also only a limited number of ways of configuring the electrical interface.

4.2.1 Data format

The interface is serial and self-clocking. That is to say that two channels of audio data are carried in a multiplexed fashion over the same communications channel, and the data is combined with a clock signal in such a way that the clock may be extracted at the receiver and used to synchronize reception. As shown in Figure 4.1, one frame of data is divided into two subframes, handling channels 1 and 2 respectively. Channels 1 and 2 may be independent mono signals or they may be the left and right channels of a stereo pair, and they are separately identified by the preamble which takes up the first four clock periods of each subframe. Samples of channels 1 and 2 are transmitted alternately and in real time, such that two subframes are transmitted within the time period of one audio sample – thus the data rate of the interface depends on the prevailing audio sampling rate.

The subframe format consists of a sync preamble, four auxiliary bits (which may be used for additional audio resolution), 20 audio sample bits in linear two's complement form, a validity bit (V), a user bit (U), a channel status bit (C) and a

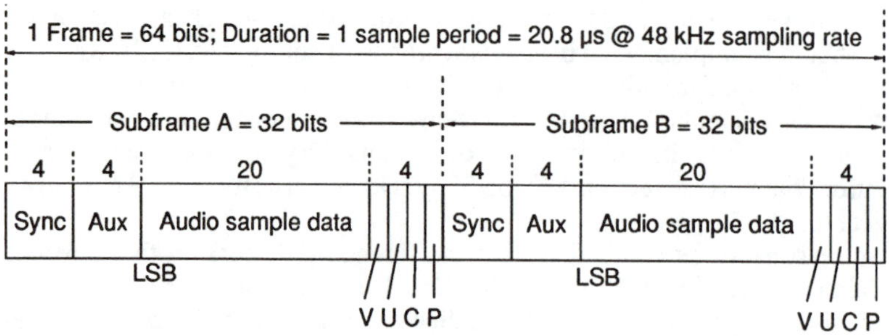

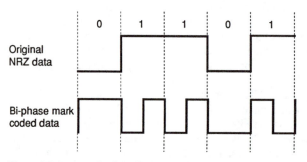

Figure 4.1 Format of the standard two-channel interface frame

Figure 4.2 An example of the bi-phase mark channel code

parity bit (P). The audio data is transmitted least significant bit first, and any unused LSBs are set to zero; thus the MSB of the audio sample, whatever the resolution, is always in the MSB position. The remaining non-audio bits are discussed in later sections.

The data is combined with a clock signal of twice the bit rate using a simple coding scheme known as *bi-phase mark*, in which a transition is caused to occur at the boundary of each bit cell (see Figure 4.2). An additional transition is also introduced in the middle of any bit cell which is set to binary state '1'. Such a scheme eliminates almost all DC content from the signal, making it possible to use transformer coupling if necessary, and also allows for phase inversion of the data signal, since it is only the transition which matters, not the direction of the transition. This channel code is the same as that used for SMPTE/EBU timecode.

As shown in Figure 4.3, there are three possible subframe preambles in time slots 1 to 4 which violate the rules of the modulation scheme in order to provide a clearly recognizable sync point when the data is decoded. These preambles cannot be confused with the data portion of the subframe. In AES3 these are called 'X', 'Y' and 'Z' preambles, but in IEC 958 they are labelled 'M', 'W' and 'B'. As the diagram shows, X and Y preambles identify subframes of channels 1 and 2 respectively, whereas the Z preamble occurs once every 192 frames in place of the X preamble in order to mark the beginning of a new channel status block (see section 4.7). Since the parity bit which ends the previous subframe is 'even parity', the transition at the start of each preamble will always be in the same (positive) direction, but a phase inverted preamble must still be decoded properly.

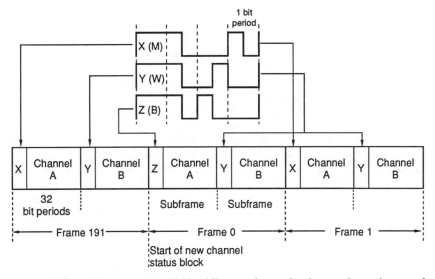

Figure 4.3 Three different preambles (X, Y and Z) are used to synchronize a receiver at the starts of subframes

The parity bit is set such that the number of ones in the subframe, excluding the preamble, is even, and thus it may be used to detect single bit errors but not correct them. Such a parity scheme cannot detect an even number of errors in the subframe, since parity would appear to be correct in this case. As discussed in section 7.3 there are more effective ways of detecting poor links than using the parity bit.

4.2.2 Audio resolution

In normal operation only the 20 bit chunk of the subframe is used for audio data, and this is adequate for most professional and consumer purposes, but the standard allows for the four auxiliary bits to be replaced by additional audio LSBs if necessary, taking the maximum resolution up to 24 bits. AES3-1992 (and other standards as they are updated) now provides a facility within the channel status data for signalling the actual number of audio bits used in the transmitted data, such that receiving equipment may adjust to decode them appropriately. This will be of considerable importance in ensuring optimum transfer of audio quality between devices of different resolutions during post-production, as discussed in section 7.2.

In consumer formats, the category code which describes the source device (see section 4.7.6) may also imply a fixed audio wordlength, since certain categories only operate at a particular resolution. The Compact Disc, for example, always uses a 16 bit wordlength.

4.2.3 Balanced electrical interface

There are two principal electrical approaches used for the standard two-channel interface: one is unbalanced and uses relatively low voltage levels, and the other is balanced and uses higher voltages. There is also a proposal for an unbalanced coaxial link for use over distances beyond 100 m (see section 4.2.6).

All the standards referring to a professional or 'broadcast use' interface specify a balanced electrical interface conforming to CCITT Rec. V.11[11]. There are distinct similarities between this and the RS-422A standard[12], and indeed RS-422 drivers and receivers are used in many cases, but they are not identical. The diagram in Figure 4.4 shows a circuit designed for better isolation and electrical balance than the basic CCITT specification, as suggested in AES3-1992. Although the transformers are not a mandatory feature of all the standards, they are advisable, since they provide true electrical isolation between devices and help to reduce electromagnetic interference problems. (Manufacturers often connect an RS-422 driver directly between the two legs of the source, which makes it balanced but not floating, or alternatively use an RS-485 driver which is a tri-state version of RS-422 giving a typical output voltage of 4 V ± 5%, going to a high impedance state when turned off.) The standards specify that the connector to be used is the conventional audio three-pin XLR (IEC 268-12), using pin 1 as the shield and pins 2 and 3 as the balanced data signal. Polarity is not really important, since the channel code is designed to allow phase inversion, although the convention is that pin 2 is '+' and pin 3 is '–'.

Although the original AES3 standard allowed for up to four receivers to be connected to one transmitter, this is now regarded as inadvisable due to the impedance mismatch which arises. Originally the standard called for the output impedance of the transmitter to be 110 ohms ± 20% over the range 0.1 to 6 MHz, and for that of receivers to be 250 ohms, but this has been changed in AES3-1992

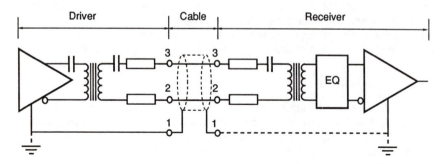

Figure 4.4 Recommended electrical circuit for use with the standard two-channel interface

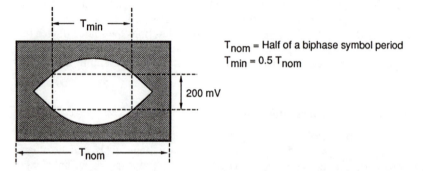

T_{nom} = Half of a biphase symbol period
$T_{min} = 0.5\ T_{nom}$

200 mV

T_{min}

T_{nom}

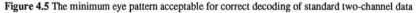

Figure 4.5 The minimum eye pattern acceptable for correct decoding of standard two-channel data

so that the receiver's impedance should now be the same as that of the transmitter and the transmission line. It is advised that only one receiver should be connected across each line, and that distribution amplifiers should be used for feeding large numbers of receivers from a single source. The cable's characteristic impedance, originally specified as between 90 and 120 ohms, is now specified as 110 ohms as well. It should be a balanced, screened pair, and although standard audio cables are often used successfully it is worth while considering cable with better controlled characteristics for large installations and long distances, in order to improve the reliability and integrity of the link (see section 7.1.1).

There is also a difference in driver voltage levels between the original and recent versions of the standard. AES3-1985, and all the other standards, specified a peak-to-peak amplitude of between 3 and 10 volts when measured across a 110 ohm resistor without the connecting cable present, but the 1992 revision now specifies it to be between 2 and 7 volts in order to conform more closely to the specifications of the RS-422 driver chips used in many systems. (RS-422A in fact specifies that receiver inputs should not be damaged by voltages of less than 12 volts.)

At the receiving end, the standards all indicate that correct decoding of the data should be possible provided that the eye pattern (see section 3.2.4) of the received data is no worse than shown in Figure 4.5. This suggests a minimum peak-to-peak amplitude of 200 mV, and allows for the toleration of a certain amount of jitter in the time domain. Without equalization the balanced interface should be capable of error-free communication over distances of at least 100 m, and often further, although this depends to some extent on the type of cable, the electromagnetic environment, the integrity of the transmission line and the quality of the data recovery in the receiver. Receivers vary quite widely in respect of their ability to lock to an unstable data signal which has suffered distortion over the link, and an interconnect which works badly with one receiver may be satisfactory with another. Devices are available which will give some idea of the quality of the received data signal, in order that the user may tell how close the link is to failure (see section 7.3).

It is possible to equalize the signal at the receiver in order to compensate for high frequency losses over long links, and the standards suggest the curve shown in Figure 4.6. It has been suggested[13], though, that as cable lengths increase the loss characteristic approaches a second order curve before problems occur, and that therefore a second order equalization characteristic is often more effective.

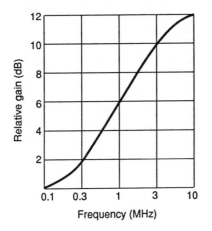

Figure 4.6 EQ characteristic recommended by the AES to improve reception in the case of long lines

4.2.4 Unbalanced electrical interface

The unbalanced electrical interface specified originally in IEC 958 and EIAJ CP-340/1201 is not a feature of professional standards such as AES3, but, whether by intention or omission, IEC 958 does not state explicitly that the unbalanced interface is for consumer use only – it simply calls it 'unbalanced line (two-wire transmission)'. The reference to a difference between consumer and professional applications comes earlier in the document, but only when discussing the structure of channel status information. For this reason, a manufacturer who puts professional data down the unbalanced interface or consumer data down the balanced interface would, it seems, still be working in conformity with the IEC 958 standard, although it would be unlikely to earn that manufacturer much in the way of thanks from users, as has been proven with one commercial DAT machine exhibiting just this feature! Operational convention has it that the unbalanced two-wire interface terminating in RCA phono connectors (which has its origins in the SPDIF format designed for CD players) is for consumer applications, and that the balanced interface terminating in XLR connectors is for professional applications. For the simple reason that this is what users expect, it would be wise for products to adhere to this convention. Interestingly, EIAJ CP-340 takes the step of noting that the unbalanced interface and the optical fibre interface apply only to Type II transmissions (consumer), although it does not say anything about the balanced interface being only for professional purposes.

The unbalanced interface is shown in Figure 4.7. IEC 958 specifies a source impedance of 75 ohms ± 20% for this interface, between 0.1 and 6 MHz, and a termination impedance of 75 ohms ± 5%. For cable runs of less than 10 m it specifies a characteristic cable impedance of 75 ohms ± 35%, but for distances over 10 m it specifies 75 ohms ± 5%. The cable is normally a standard audio coaxial cable, and this interface is typically used for interconnecting consumer equipment over the sorts of distances involved in hi-fi systems. It does not specify a maximum length over which communication may be expected to be successful, and it does not give an eye pattern limit for correct decoding, although it does specify a minimum peak-to-peak input voltage at the receiver of 200 mV. A significant difference between this interface and the balanced interface is that the source signal amplitude should

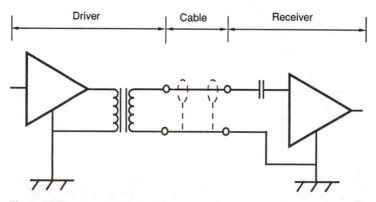

Figure 4.7 The consumer electrical interface

be only 0.5 V ± 20%, peak-to-peak, which is much lower than the balanced interface. It should be noted, though, that video-type 75 ohm coaxial cable exhibits very low losses below about 10 MHz, and thus one might expect to be able to cover significant distances without the signal level falling below the minimum specified.

It used to be said by some that since the unbalanced interface was a coaxial transmission line with well-controlled impedances it formed a better link than the balanced interface, but this was always offset by the advantages of a balanced line in rejecting interference, and the higher voltages used in the balanced interface. Now that the balanced interface specifies source and termination impedances to be the same, requires point-to-point connection, and recommends 110 ohm cable (rather than anything between 90 and 120 ohms), the balanced interface has the benefits of a good transmission line as well as its other advantages.

Any revision to the standard which might introduce the possibility of a coaxial interface using BNC connectors for professional applications would be likely to use higher voltage levels than the IEC 958 unbalanced interface, in order to be compatible with video-type circuits and distribution amplifiers operating at around the 1 volt level.

4.2.5 Optical interface

An optical interface was introduced as a possibility in IEC 958 but was left 'under consideration'. It is specified more explicitly in EIAJ CP-340 (or CP-1201) as applying only to Type II data, and consisting of a transmitter with a wavelength of 660 nm ± 30 nm and a power of between –15 and –21 dBm. Receivers should still correctly interpret the data when the optical input power is –27 dBm. The connector indicated conforms to the specification laid out in EIAJ RCZ-6901. The standard acknowledges that experience with optical fibres at the time of writing was limited, and thus less information on implementation was presented than for the electrical interfaces.

Typically the optical interface is found in consumer equipment such as DAT recorders, CD players, stand-alone convertors, and amplifiers with built-in D/A convertors. It usually makes use of an LED transmitter (see section 1.7) and a fibre optic cable, connected to a photodetector in the receiver. The 'TOSLink' style of fibre optic interface is popular in consumer equipment, and is driven from a TTL level (0–5 volt) unbalanced source, with a data format identical to that used with the electrical interface. The advantages of optical links in rejecting interference have already been stated in section 1.7, and it is possible that such links may begin to find greater favour in professional applications in the future, but there are dangers in using cheap optical interfaces since their limited bandwidths and high dispersion may actually result in a poorer transmission channel than a normal electrical interface, resulting in a high degree of timing instability in the positions of data transitions (see Chapter 6). For a comprehensive introduction to fibre optics in audio the reader is referred to Ajemian and Grundy[14].

4.2.6 Coaxial interface

As mentioned above, it is proposed also to introduce a coaxial version of the professional AES3 interface, provisionally entitled AES-3ID, which makes use of 75 ohm video-style coaxial cable to carry digital audio signals over distances of around 1000 m. A signal level similar to that of a video signal (1 volt) is envisaged, although the signal is not formatted to look like a video signal – it is still basically the same bi-phase mark channel code of AES3. Although the standard is not finalized at the time of writing, it is hoped that easy adaption will be possible between the balanced form of AES3-1992 and this coaxial form: indeed a number of manufacturers already make balanced-to-coax adaptors for just this purpose (although the voltage level will be much lower after transformation from 1 volt/75 ohms back to 110 ohms than might be expected from a standard AES3 balanced output stage). The advantages of this interface would be the ease of distribution of audio within a television studio environment, using video distribution amplifiers and cabling, and improved electromagnetic radiation characteristics when compared with the balanced twisted pair, as discussed in Rorden and Graham[15].

Preliminary tests of such an interface have been successful, showing in one test that equalized video lines could carry an AES-format digital audio signal over a distance of more than 800 miles (1300 km) without any noticeable corruption.

4.2.7 Multipin connector

A multipin connector version of AES3 is also proposed, although not standardized at the time of writing. Such an interface would allow multiple channel interfacing without resorting to the MADI standard (see section 4.8), and could be a lower cost solution than MADI in cases of a smaller number of channels than 56. One strong candidate for this version is designed to carry 16 channels of audio data on a single 50 pin D-type connector. This is likely to be included in the forthcoming Engineering Guidelines on the interface.

4.3 Sampling rate related to data rate

The standard two-channel interface is designed to accommodate digital audio signals with sampling rates normally between 32 and 48 kHz, with a margin of ±12.5% to allow for varispeed operations. Since the interface carries audio data in real time, transferring two audio samples (channel 1 and channel 2) in the time of one sampling period, the data rate of the interface depends on the audio sampling rate. It is in fact 64 times the sampling rate, since there are 64 bits in a frame (two subframes). At a sampling rate of 48 kHz the data rate is 64 times 48 000, which is 3.072 Mb/s, whereas at 32 kHz it is only 2.048 Mb/s. If the source is varispeeded by a certain percentage then the data rate will change by the same percentage, and although it can usually be tracked by a receiver this presents problems in a system where all devices must be locked to a common, fixed sampling frequency reference (see Chapter 6), since the receiver may not change its sampling rate to follow a varispeeded source.

Further considerations relate to the timing accuracy of data on the interface, which is specified to be within certain limits depending on whether the application is consumer or professional. Questions of timing accuracy which arise in interface data signals should be considered together with questions of sampling frequency accuracy and effects on sound quality in convertors, although they are strictly separate matters, and these criteria are also closely related to the topic of synchronization in digital audio interfacing. For these reasons further discussion of the matter will be postponed until Chapter 6.

4.4 Auxiliary data in the standard two-channel interface

As stated earlier, the 4 bits of auxiliary data in each subframe may be used for additional LSBs of audio resolution, if more than 20 bits of audio data are needed per sample. Alternatively the auxiliary data may be used to carry information associated with the audio channel, and until recently no consensus or standard existed over how these bits should be used. In many items of equipment manufactured to date they remain unused.

However, it was proposed to the CCIR in 1987 that the aux bits would prove useful for a good voice quality channel which could be used for coordination (talkback) purposes in broadcasting[16]. Typically in a radio broadcast studio the programme source (say a studio) sends a stereo programme to a destination (say a continuity suite) along with a good voice quality link for coordination purposes (see Figure 4.8). A feed of cue programme (normally mono), together with a coordination channel and perhaps additional data, is returned from the destination. It was proposed that in digital studio environments all of these signals could be carried over a single standard two-channel interface in each direction by sampling the coordination voice channel at exactly one-third of the main audio's sampling frequency and coding it linearly at 12 bits per sample, resulting in a data rate exactly one-fifth that of the main audio channel. (Main audio channel @ 48 kHz, 20 bits; Data rate = 960 000 bits per second; Coordination channel @ 16 kHz, 12 bits; Data rate = 192 000 bits per second.) At such a data rate, a main sampling rate of 48 kHz would allow for a coordination channel bandwidth of about 7 kHz.

Capacity exists in the two-channel interface for two 12 bit coordination samples, one in channel 1's subframe ('A' coordination) and one in channel 2's ('B' coordination). They are inserted 4 bits at a time, as shown in Figure 4.9, with the four LSBs of the 'A' signal going into the first aux word in the block (designated by the Z preamble), followed by the four LSBs of the 'B' signal in the next subframe, and so on for three frames, whereon all 12 bits will have been transmitted for each signal. The process then starts over again. The Z preamble thus acts as a sync point for the coordination channels, and the sampling frequencies of the coordination channels are locked to that of the main audio.

This arrangement proves very satisfactory because the correct talkback always accompanies a programme signal, the cue programme may be in stereo, and only two hard links are necessary. The user bit channel of the interface can be used to carry any additional control data. In the 1990 revision of CCIR Rec. 647[5] this modification was taken on board, and forms part of that standard. It was also adopted by the EBU in the 1991 revision of Tech. 3250-E, and appears as an annex to AES3-1992 (for information purposes only). In order to indicate that the aux bits are being

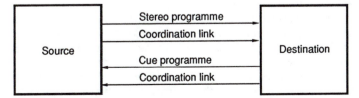

Figure 4.8 In broadcasting a coordination link often accompanies the main programme, and cue programme is fed back to the source, also with coordination

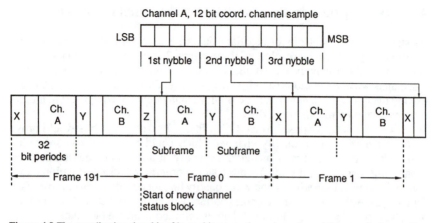

Figure 4.9 The coordination signal is of lower bit rate to the main audio and thus may be inserted in the auxiliary nybble of the interface subframe, taking three subframes per coordination sample

used for this purpose, byte 2 of the channel status word (see section 4.7.1) is used. The first 3 bits of this byte are set to '010' when a channel's aux bits carry a coordination signal of this type.

4.5 The validity (V) bit

The application and value of the validity bit in each subframe have been the subject of considerable conjecture during the formulation of standards. Originally the V bit was designed to indicate whether the audio sample in that subframe was 'valid' or 'reliable', 'secure and error free' – in other words, to show either whether it contained valid audio (rather than something else, or nothing), or if it was in error. It was set to '0' if the sample was reliable, and '1' if unreliable (so really it is an 'invalidity' flag). Since it is only a single bit, there is no opportunity for signalling the extent or severity of the error. What has never been clear is what devices should do in the case of an invalid sample, and this is largely left up to the manufacturer. The most common use for the V bit is to signal errors which occurred in the transmitting device, such as when an uncorrectable error is encountered when replaying a recording for example. But not all devices treat this in the same way, since some signal any offtape CRC error whether it was corrected or not, whereas others only set the V bit if the error was uncorrectable, resulting in interpolation or even muting in the convertors of most systems (this seems the more appropriate solution). But this is not the only use, since interactive CD (CD-I) players, for example, use the V bit to indicate that the audio data part of the subframe has been replaced by non-audio data, since otherwise there would be a potential delay of up to 1 block (192 frames) before the receiving device realized that the channel use had changed from audio to non-audio (this is normally signalled in channel status which is only updated once per block).

In AES3-1992 the description of this bit has changed somewhat and it now indicates whether the audio information in the subframe is 'suitable for conversion to an analog signal', which may or may not amount to the same thing as before. (A binary '1' still represents the error state.) For example, the V bit may be set if an uncorrectable error arises on a DAT machine which would normally result in error concealment by interpolation rather than muting at the output. Any subsequent device receiving audio data from this machine over the digital interface would see the V bit set true and, if interpreting it literally, would assume that the audio was unsuitable for conversion and mute its output, yet the user might still want to hear it, assuming that the interpolation sounds better than a mute! Receivers vary in this respect, and there are some which always mute on seeing the V bit set true. A further and potentially more serious problem, highlighted by Finger[2], is that a recording device will usually store incoming audio data but has no means of storing the validity flag. In such a case the replayed audio would then be transmitted with no indication of invalidity, even if recorded errors existed.

Another problem arises in devices which simply process the data signal, such as sample rate convertors or interface processors such as those described in Chapter 7. Should such devices take any action in the case of invalid samples, or should they simply pass the data through untouched? One such device takes the approach of carrying through the V bit state and holding the last sample value in the case of an error, but what happens when two digital signals are to be mixed together, one which is erroneous and the other not? In such a case it is difficult to decide whether the single mixed data stream should be valid or invalid and there are no clear guidelines on the matter (indeed there is no 'catch-all' solution to such a problem).

It is intended that the Engineering Guidelines soon to be issued as an accompaniment to AES3 will address this matter further, and it is probable that these will contain advice on specific situations, but one should not expect magic solutions to this problem since there is still only one validity bit and any alternative approach would still have to apply on a sample-by-sample basis, rather than at block level (in which case perhaps channel status could be used). The most likely outcome is that the interpretation of the validity bit will be left 'application dependent', allowing manufacturers to deal with it depending on how the importance of errors is viewed in the particular application.

4.6 The user (U) channel

The U bit of each subframe has a multiplicity of uses, many of which have remained hidden from the user of commercial equipment, such as the carrying of text, subcode, and other non-audio data. It is most widely used in consumer equipment, and there is now a rather complicated AES standard for its use in professional applications. There is also a Philips method for inserting data into the user channel, called ITTS (Interactive Text Transmission System), on which the CD and DCC systems rely for the transference of subcode and other non-audio data over the consumer interface. The U bit is only a single bit in each subframe, allowing a user channel to accompany each audio channel, and its definition in the various standards is normally 'for any other information'. The user bits are not normally aggregated over the same block length as channel status data (192 subframes), although they may be, but are often aggregated over different block lengths depending on the application, or may simply be used as individual flags. Many devices, especially professional ones, do not use them at all, although this may change in the future.

In the following sections a number of the most common applications for user data are outlined, although the standard does not prohibit users or manufacturers using

Table 4.1 Indication of user bits format in channel status byte 1

Bits 4–7	User bits format
0000	Default, no user information
0001	192 bit block structure
0010	AES18 (see section 4.6.1)
0011	User defined

this capacity for alternative purposes. AES3-1992 signals the use of the user bits in byte 1, bits 4–7 of channel status as shown in Table 4.1.

(It may be noted that the terminology used to describe the bit number of a message in the user channel can be confusing, depending on whether the message is considered as running 'MSB to LSB' or vice versa. We shall refer to the first transmitted bit of a message byte as 'bit 0', but some documents refer to this as 'bit 7'.)

4.6.1 HDLC packet scheme (AES18-1992)

Unlike channel status data, user data may consist of a wide variety of different message types, and the AES working group on 'Labels' decided that the best approach to the problem for professional users was to allow the user channel to be handled in a 'free format', such that its maximum capacity of 48 kbit/s could be shared between applications, with user data multiplexed into 'packets' of information which would share the interface. The history of this goes back to 1986, and a proposal by Roger Lagadec of Sony, in which it was suggested that user data was very different to channel status data and would not suffer the same block structure, requiring a more flexible approach in which some messages could be sent once with minimal delay, whereas others might be repeated at regular time intervals, and yet others might have to be time-specific. It required that the data rate in the user channel be independent of the audio sampling rate, whereas the actual rate of user bits depends on the interface frame rate, and thus on the sampling rate.

It is not necessary to document the whole history here, except to say that the direction of the work was influenced considerably by proposals from TDF (Télédiffusion de France) and others, well documented by Komly and Viallelieille[17], suggesting the use of an asynchronous frame format already well established in the telecommunications and computer industries called HDLC (High-level Data Link Control). This is an internationally standardized way of transferring data at a bit-oriented level around networks (ISO 3309-2)[18], and there are a number of commercial chips available which do the job of inserting data into the correct packet structure. The working group finally recommended a structure for carrying user data in AES18-1992[19], and a useful commentary on this may be found in Nunn[20]. If this particular way of treating the user bits is implemented then it is indicated in byte 1 of the channel status information, bits 4–7, as shown in Table 4.1.

Although this is a flexible and versatile way of treating the user bit channel it might be criticized for being overcomplicated by many people, and it lacks a defined message structure[21], leaving it up to the user to build his or her own applications around the protocol. It could be said that this is the role of a system designed to handle data of virtually any sort, but it remains to be seen whether it will be taken up by many manufacturers or systems designers, given that the concepts involved are more familiar to computer network designers than to audio engineers. It should be stressed that currently this approach is not part of the consumer format.

Among the key features of this standard are that the data rate of the user channel can be kept constant over a defined range of sampling rates (but only between 42 and 54 kHz in AES18), that a precise timing relationship can be maintained between audio and user data, that time-critical data may be transmitted within a specified and guaranteed period, and that the channel may be used simultaneously by a number of users. User data to be transmitted is formed into packets which are preceded by a header containing the address of the destination, and the packet is then inserted into the user data stream as soon as there is room. In order to ensure that the user data rate remains constant down to an audio sampling rate of 42 kHz (which is 48 kHz minus 12.5%) extra packing bits are added at the end of each block of packets which can be disposed of as the sampling frequency is lowered. At audio sampling rates below 42 kHz the data rate will be lower, and thus some information would be lost if 48 kHz data were to be sample rate converted to, say, 44.1 or 32 kHz, but it is expected that some form of data management would be implemented to ensure that important data gets the highest priority in these circumstances.

Data is formatted at a number of levels before being transmitted over the interface, starting at the highest level – the 'application level' – and ending at the lowest level – the 'physical level' – at which the data is actually inserted bit by bit into the audio interface subframe structure. It is not intended to cover the process by which this is achieved here, since this would constitute needless repetition of available documentation. What is important is some commentary on the handling of different types of message, particularly time-specific messages, and on the insertion of additional messages at later points in the interface chain.

AES18 allows for the handling of time-specific messages by formatting the user data packets into blocks, normally of fixed but definable length, and repeating these at a user-definable rate which can be set to correspond to time intervals pertinent in the application concerned. An optional 'system packet' may also be transmitted at block intervals which may contain timecode data amongst other things, and sets priorities for different types of message which may have more or less urgency. It recommends some useful repetition rates of blocks, which correspond to the timing intervals of frames in audio and video applications, as shown in Table 4.2.

In some applications variable block lengths may be necessary, such as when using 48 kHz audio with NTSC video (which runs at 29.97 fps) where there are not an integer number of audio samples per video frame.

In order to allow for the insertion of messages of varying importance at different points in the system, the standard sets down comprehensive rules governing the way in which messages should be prioritized. The maximum delay involved in inserting a packet of data depends on its priority (from 0 to 3), and the block length involved. The highest priority packet (level 3) may be inserted once per block, and as the

Table 4.2 Some useful repetition rates of blocks

Blocks per second	Duration (ms)	Application
24	41.67	Film
25	40	PAL, SECAM video or 50 frame per second (fps) HDTV
29.97	33.37	NTSC video
30	33.33	60 fps HDTV
33.33	30	DAT

priority is decreased the packets are inserted only once per so many blocks. Since the shortest practical block length is 10 ms, this is the minimum delay one might anticipate.

4.6.2 Applications of the user bit in Compact Disc systems

Compact Disc players having consumer format digital outputs normally place subcode data in the user bits. This is in addition to the control bits of the Q-channel subcode from the CD which is transmitted within channel status (see section 4.7.9), and is the only full specification for user bits implementation contained within the original IEC 958 document, as Annex A, although later and forthcoming amendments to IEC 958 specify detailed applications with other types of equipment such as DAT (Digital Audio Tape), DCC (Digital Compact Cassette) and MD (MiniDisc). EIAJ CP-340, however, also details the basic application of user bits with DAT machines.

In the CD application the user bits for the left and right subframes are treated as one channel, and the Q to W subcode data is multiplexed between them (the P flag is not transmitted since it only represents positioning information for the CD transport). The subcode data block in CD is built up over 1176 samples, formed into CD sync blocks of 12 samples each (making 98 subcode symbols, which include 2 symbols for block sync). There are eight subcode bits in each of these sync blocks (P–W), but only seven of them are transmitted (Q–W) over the interface. Since the subcode data rate from CD is therefore lower than the user bit rate, zeros are used as packing between the groups of subcode bits. The number of zeros is variable, principally to allow for variable speed replay to ±25%. As shown in Figure 4.10, the subcode block begins with a minimum of 16 zeros, followed by a start bit (a binary '1' which some documents call 'P' although this might be misleading since it is

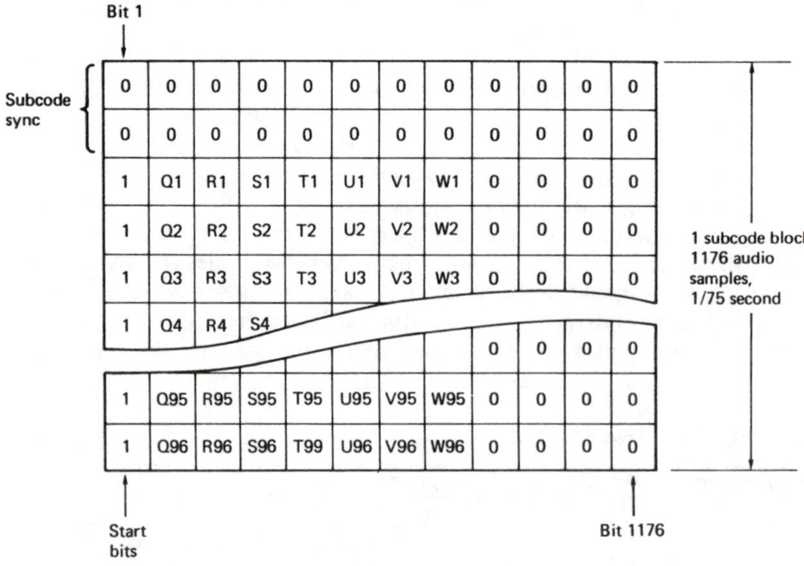

Figure 4.10 An example of user bits formatting in the CD system

always a '1' and thus contains no additional information). There then follow seven subcode bits (Q1 to W1), after which there may be up to eight zeros before the next start bit and the next 7 bits of subcode data. Only four packing zeros are shown in this diagram. (In this respect the text of IEC 958 (1989), Annex A, appears to be incorrect, although the diagram is correct, since the text specifies that the distance between start bits may be minimum 8 bits and maximum 16 bits.) This pattern is repeated 98 times, after which a new sync pattern of at least 16 zeros is expected.

The Q data in the subcode stream can be used to identify track starts and ends, amongst other things (see the full CD specification in IEC 908), and thus it is useful when transferring CDs to DAT or vice versa (for professional purposes, of course), or from a CD player to a CD recorder, since the audio data and the track IDs are duplicated together and the copy is a true clone of the original. Between CD machines there is usually little problem in copying subcode data, since the two machines are of the same format, but between CD and DAT a special processor unit is normally required to convert DAT track IDs to CD track IDs or vice versa, and there are occasional discrepancies. Since the P flag is not transferred over the interface the copy may only rely on Q subcode information, and there is usually a gap between the start of the P flag on the CD and the Q subcode track number increment. Some CD players increment the track number on their own displays at the start of the P flag and then count down to the true track start using the Q data, whereas a copy of such a recording would only increment the track number at the true track start. There is also occasionally a small delay in the assertion of the track start flag on DAT recordings, due to the automatic start ID facility used in many machines which writes a new start ID when the audio level rises above a certain point, which may sometimes be compensated for in the transfer.

4.6.3 Applications of the user bit in DAT systems

As with the CD, the consumer interface on DAT machines also carries some additional information in the user bits. The first edition of IEC 958 is now technically incorrect in Annex C, since it suggests that subcode data will be carried in the four auxiliary bits rather than the user bits. Annex C of IEC 958 is now being modified to signify subcode in user bits, with nothing in the aux bits. Considering the subcode information which could be sent in the user bits the actual implementation is incredibly crude, since it simply indicates the presence or lack of start and skip (shortening) IDs on the tape. This approach was in fact inherent in the DAT design standard right from the start[22].

As shown in Figure 4.11, sync, start ID and skip ID are transmitted over the interface with relation to the DAT frame rate of 33.33 frames per second. As with CD, the user bits of the left and right channel subframes are considered together. The sync ID is transmitted once per frame by setting the user bit of the first left channel sample (L_0) of that frame true – this simply indicates where the frame begins and could be used for crude synchronization of two machines. (In other words, the user bit of the interface subframe corresponding to the first sample of the DAT frame is always set true.) When a start ID is present on the tape the user bit of the following interface subframe (which is the first right channel sample of the DAT frame, or R_0) is also set true, and this lasts for 300 ± 30 frames, or about 10 seconds (the same duration as the start ID information on the tape). When a skip ID is encountered in normal play mode (without actually skipping) the user bit of the next left channel subframe (L_1) is set true, and this is repeated for 33 ± 3 frames, or about 1 second.

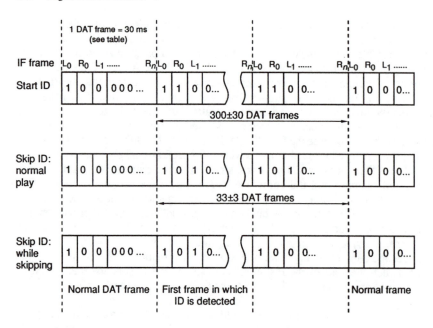

L_0 user bit is SYNC and goes true every DAT frame (every n +1 IF frames)

R_0 user bit is Start ID and is true for 300±30 frames (i.e. for duration of Start ID on tape)

L_1 user bit is Skip ID and is true for 33±3 frames when not skipping (just detecting a Skip ID on tape), but is true only for 1 frame at the start of an actual skip (in skip play)

Sampling rate	Digital IF frames per DAT frame	n
32 kHz	960	959
32 kHz (LP)	1920	1919
44.1 kHz	1323	1322
48 kHz	1440	1439

Figure 4.11 Signalling of DAT start and skip IDs in user bits (user bits only shown)

When the DAT machine is programmed to act on skip IDs it will skip to the next start ID, and in this case the user bit of the L_1 frame is set true only once – in the first frame that it is encountered. All the other user bits are set to zero.

The number of samples corresponding to a DAT frame depend on the sampling rate, and therefore this dictates the distance between sync, start and skip IDs in the user bits of the interface. At 48 kHz there are 1440 left and right samples per frame, making 2880 subframes between the occurrence of these user bits. At 44.1 kHz this gap is reduced to 2646 words, and in the 32 kHz long play mode found on some players it is 3840 words.

4.6.4 Application of the user bit in DCC systems

Two modes for user data are currently allowed for in the new digital compact cassette (DCC) system. These are called 'marker mode', which is mandatory, and 'extended mode', which is optional. The data to be transferred in the user bits is formatted into messages, and these messages are made up of one or more information units (IUs) which are 1 byte long and always begin with a binary 1 (the start bit). The second bit of the first IU of a message is the 'mode bit', and this is set to '0' for marker mode and '1' for extended mode. IUs within one message may be separated by up to eight zero bits (rather like the Compact Disc subcode), and messages must be separated by more than eight zero bits. User bit data is multiplexed between left and right channel subframes as with the CD.

A marker mode message is only 1 byte long and is transmitted once per DCC frame, with the format shown in Figure 4.12, and contains basic control information as follows:

Bit 2 (LAB): Set to 1 in the presence of a start ID on tape
Bit 3 (SH):   Set to 1 when a skip ID is encountered
Bit 4 (FAD): Set to 1 to program a fade down or up, to occur at the mute flag
Bit 5 (MUT): Set to 1 when a mute or fade starts (remains muted until set to 0)
Bit 6 (STP): Set to 1 to indicate stop mode or no available audio signal
Bit 7 (SCM): Set to 1 in the presence of a sector marker on tape

An extended mode message may consist of several IUs. The first byte of an extended mode message has the format shown in Figure 4.13, and uses the six active bits of the IU to indicate the number of the message to be transmitted. The role of these messages is defined within the DCC standard, and the reader is referred to the Philips ITTS (Interactive Text Transmission System) documentation[23] for further information on the format of text data within extended messages.

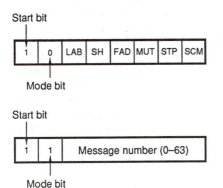

Figure 4.12 Format of marker mode message in DCC

Figure 4.13 Format of the header byte of an extended mode message in DCC

4.6.5 Application of the user bits in MiniDisc systems

Although no published documentation could be obtained by the authors concerning the implementation in MiniDisc, it appears from preliminary information that the first MD products will use an implementation similar to that of the Compact Disc system. Q subcode data is contained in the user bit channel using the same format as for CD (see above).

4.7 Channel status data

Channel status data (represented by the C bit in each subframe) is normally the most problematic area within the standard two-channel interface. It is here that a number of common incompatibilities arise between devices, and it is here that the main differences exist between professional and consumer formats, since the usage of channel status in consumer and professional equipment is almost entirely different. In this section the principles of channel status usage will be explained, together with an introduction to potential problem areas, although discussion of practical situations is largely reserved until Chapter 7.

4.7.1 Format of channel status

Although there is only one channel status (C) bit in each subframe, these are aggregated over a period of time to form a large data word which contains information about the audio signal being transmitted. The two audio channels theoretically have independent channel status data, although commonly the information is identical, since most applications are for stereo audio. Starting with the frame containing the Z preamble, channel status bits are collected for 192 frames (called a channel status block), resulting in 192 channel status bits for each channel. This long word is subdivided into 24 bytes, each bit of which has a designated function, but the function of these bits depends on whether the application is consumer or professional. The channel status information is updated at block rate, which is 4 ms at a sampling rate of 48 kHz, and longer *pro rata* at other sampling rates.

Standards such as AES, EBU and CCIR only cover the professional application of channel status, but the EIAJ, IEC and British standards all include a section on consumer applications as well. In fact it is the consumer applications of channel status that are always being updated in these standards, as new consumer equipment appears with different facilities, whereas the professional data remains relatively unchanged. One exception to this is in the recent revision to AES3 (1992) in which the format of professional channel status data has been extended and explained more carefully, with the intention of setting down more clearly what devices must do with this data if they are to conform properly with the standard. These revisions to the professional use of channel status will gradually find their way through to the other standards, as indeed they are already doing, and will hopefully ensure that future professional devices are more compatible with each other, although it will not help the situation with older equipment. The problem of incompatibility in channel status data is covered further in Chapter 7.

4.7.2 Professional and consumer usage compared

It is both a blessing and a curse that the formats of professional and consumer data are so similar. It is a blessing because the user may often wish to transfer material from one to the other and the similarity would appear to make this possible (see section 7.1), but it is a curse because the usage of channel status is so different between the two that the potential for problems is very high, due to the likely misinterpretation of this information by a receiver of the other format. (There is no technical reason, though, why a device should not be designed to interpret both formats correctly.) The first bit of the channel status block indicates whether the

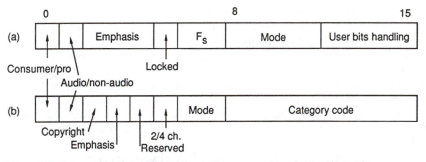

Figure 4.14 Comparison of (a) professional and (b) consumer channel status bits 1–16

usage is consumer (0) or professional (1), and this bit should be interpreted strictly in order to avoid difficulties. (It should be noted that before the consumer format was standardized by the IEC the first bit of the consumer channel status data actually represented 'four-channel mode', not 'consumer/professional', and some early devices may still exhibit this feature.)

Figure 4.14 shows a graphical comparison of the beginnings of the channel status blocks in professional and consumer modes. Clearly bit 0 is the same and indicates the usage, and bit 1 is also the same, indicating audio (0) or non-audio (1) usage of the interface, but here the similarity stops since in the professional version there are 3 bits (bits 2, 3 and 4) used to signal pre-emphasis, whereas in the consumer version there are only 2 (bits 3 and 4) for emphasis (although only one is actually used, since bit 4 is reserved for future use), with bit 2 being used to signify copyright material. Already there is room for incompatibility since a professional device trying to interpret a consumer signal or vice versa could confuse copy protection states with emphasis states, even though IEC 958 says that bits 0–4 are designed to allow a professional device to accept a consumer signal. Since copy protection is not normally an issue in professional applications, there is no provision for signalling it in the professional interface. A copyrighted consumer signal with pre-emphasis would have bits 2, 3 and 4 set to '010', which is a disallowed state in IEC 958's professional specification, and if it was accepted would probably be interpreted as 'emphasis not indicated' and default to no emphasis.

4.7.3 Professional usage

Figure 4.15 shows the format of the professional channel status block, and Figures 4.16(a) and (b) show the functions of the first and second bytes in more detail. They indicate basic things about the nature of the source, such as its sampling frequency (a root of problems, as discussed in section 7.1.1), its pre-emphasis, the mode of operation (mono, stereo, user defined, etc.) and the way in which the user bits are handled (see section 4.6 and Table 4.1).

Byte 2 will become increasingly useful as systems with resolutions greater than 16 bits appear, since it allows the source to indicate the number of bits actually used for audio resolution in the main part of the subframe. This would allow receiving devices to adapt their signal processing in order to handle the incoming signal resolution appropriately, such as when a 16 bit device receives a 20 bit signal, perhaps re-dithering the audio at the appropriate level for the new resolution in order to avoid distortion. This byte was less comprehensively used in the original version of AES3, only indicating whether the audio wordlength was maximum 20 bits or 24 bits (in other words, whether the aux bits were available for other purposes or not), but in AES3-1992 it allows for greater detail. Figure 4.17 shows how byte 2 is split up, with the first 3 bits describing the use of the auxiliary bits (see section 4.4), and Table 4.3 shows how bits 3–5 are used to indicate audio wordlength. Bits 6 and 7 are still reserved for future use.

It is important to remember that, no matter what the audio resolution, the MSB of the audio sample should always be placed in the MSB position of the interface subframe.

The first 2 bits of byte 4 of professional channel status are used for indicating whether the signal can be used as a sampling frequency reference, according to the AES11 standard on synchronization. In the '00' state these bits indicate that the signal is not a reference, whilst the '01' state is used to represent a Grade 1 reference and the '10' state is used to represent a Grade 2 reference. This topic is discussed in greater detail in Chapter 6.

Byte

0	Basic control data (see Figure 4.16)
1	Mode and user bit management (see Figure 4.16)
2	Audio wordlength (see Figure 4.17)
3	Vectored target from byte 1 (reserved for multichannel applications)
4	AES11 sync ref. identification (bits 0–1), otherwise reserved
5	Reserved
6	
7	Source identification (4 bytes of 7 bit ASCII, no parity)
8	
9	
10	
11	Destination identification (4 bytes of 7 bit ASCII, no parity)
12	
13	
14	
15	Local sample address code (32 bit binary)
16	
17	
18	
19	Time-of-day sample address code (32 bit binary)
20	
21	
22	Channel status reliability flags (see Figure 4.18)
23	CRCC

Figure 4.15 Overall format of the professional channel status block

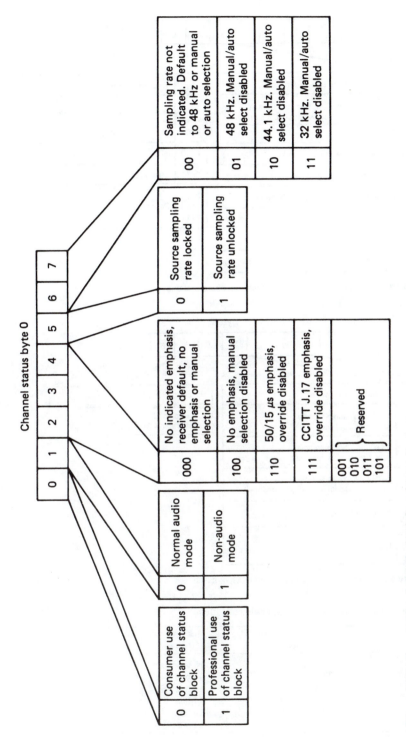

Figure 4.16(a) Format of byte 0 of professional channel status

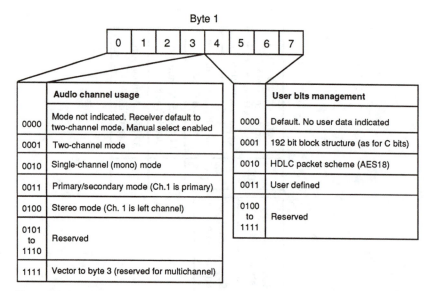

Figure 4.16(b) Format of byte 1 of professional channel status

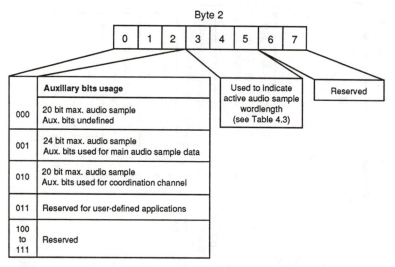

Figure 4.17 Format of byte 2 of professional channel status

Table 4.3 Use of byte 2 to represent audio resolution

Bits states 3 4 5	Audio wordlength (24 bit mode)	Audio wordlength (20 bit mode)
0 0 0	Not indicated	Not indicated
0 0 1	23 bits	19 bits
0 1 0	22 bits	18 bits
0 1 1	21 bits	17 bits
1 0 0	20 bits	16 bits
1 0 1	24 bits	20 bits

In bytes 6 to 13 it is possible to transmit information concerning the source and destination of the signal in the ASCII text format used widely in information technology (see ISO 646). ASCII characters are normally 7 bits long (although extended character sets exist which use 8 bits), and can represent alphanumeric information. The eighth bit is often used as a parity bit in telecommunications, but it is not used in this application, being set to zero. (AES3-1985 and IEC 958 specified odd parity, but this was changed in AES3-1992 to no parity.) Some ASCII characters are non-printing symbols called 'control characters' (the first 31 (hex 01–1F) and the last one (hex 7F)) and these are not permitted in this application either. Using this part of channel status the user can 'stamp' the audio signal with a four-character label to indicate the name of the source (bytes 6–9), and the same for the destination (bytes 10–13). It is possible to use destination labelling in automatic routers in order that a signal may control its own routing. The format for these ASCII messages is LSB first, and with the first character of each message in bytes 6 and 10 respectively.

Bytes 14–21 carry what are called 'sample address codes', which are a form of timecode but counting in audio samples rather than hours, minutes, seconds and frames. Bytes 14–17 are a so-called 'local sample address' which can be used rather like a tape counter to indicate progress through a recording from an arbitrary start point, and bytes 18–21 are a time-of-day code indicating the number of samples elapsed since midnight. Usually this is the time since the device was reset or turned on, because most devices do not have the facility for resetting the sample address code to time of day, but AES3-1992 contains a note to state that it should be the time of day which was laid down during the original source encoding of the signal, and should not be changed thereafter, implying that it should be derived from offtape timecode if that exists.

The 4 bytes for each sample address code are treated as a 32 bit number, with the LSB sent first; 32 bits allows for a day of 4 294 967 296 samples, which represents just over 24 hours at a sampling rate of 48 kHz. (At the maximum sampling frequency normally allowed for over the interface (54 kHz) 32 bits are not quite enough to represent a whole day's worth of samples, allowing a count of just over 22 hours, but this sampling frequency is normally only used as an upward varispeed of 48 kHz and thus is a non-real-time situation in any case.) If the sampling frequency is known then it is a straightforward matter to convert the sample address to a time of day, although other methods are being considered for allowing specific SMPTE/EBU timecode information to be encoded within the user bits of the interface. The sample address code is updated once per channel status block; thus it is incremented in steps of 4 ms at 48 kHz, and represents the sample address of the first sample of that block.

Byte 22 is used to indicate whether the data contained in certain channel status bytes is reliable, by setting the appropriate bit to '1' in the case of unreliable information, as shown in Table 4.4.

Table 4.4 Use of byte 22 to indicate reliability of channel status data bytes

Bits	Bytes reliable
0–3	Reserved
4	Bytes 0–5
5	Bytes 6–13
6	Bytes 14–17
7	Bytes 18–21

The last byte of the channel status block (byte 23) is a CRC (Cyclic Redundancy Check) designed to detect errors *in channel status information only*, and a simple serial method of producing CRC information is given in AES3-1992. (Most modern AES/EBU interface chips generate channel status CRC automatically.) The CRC byte was not made mandatory in AES3-1985, and thus a number of systems have not implemented it, but in AES3-1992 it is mandatory in the 'standard' implementation of the interface (although there is a 'minimum' mode in which it can be left out). IEC 958 (1989) also makes things difficult since it says that if channel status is implemented then all data in bytes 0 and 1 must be transmitted, but it does not insist on CRC. The presence or lack of the CRC byte in different devices is a root of incompatibility in many cases, since devices expecting to see it may refuse to interpret channel status data which does not contain a CRC byte because they will assume that it is always in error.

Any bits in channel status which are either not used or reserved should be set to the default state of binary '0' – another important factor in ensuring compatibility between devices.

4.7.4 Levels of professional channel status implementation

AES3-1985 was rather unclear as to what manufacturers should do with channel status if they were not implementing certain features, and indeed receivers fell between the two extremes of either interpreting the standard so literally that they expected to see every single bit set 'correctly' before they would work, or else interpreting it so loosely that virtually anything would work, often when it should not. (To be fair it is true that most of the time communication worked without a problem.) In the 1992 revision it was decided to recommend three levels of implementation at the transmitter end, encouraging manufacturers to state the level at which they were working. These levels have been called 'minimum', 'standard' and 'enhanced'. It is intended that at all levels the rest of the frame should be correctly encoded according to the standard.

At the minimum level channel status bits are all set to zero except for the first bit which should be set to signify professional usage. Such an implementation would allow 'belt and braces' communication in most cases, but leaves room for many problems since CRC is not included and neither is any indication of pre-emphasis or sampling frequency. It is intended in such a case that the receiver should set itself to the default conditions (48 kHz, no emphasis, two-channel mode) but allow manual override.

At the standard level the transmitter is expected to implement bytes 0 to 2 and 23 of channel status. This then allows for all the information about the source signal and use of different bits in the frame to be signalled, as well as including the CRC. The standard implementation differs from that stated in IEC 958 (1989) in which a minimum of bytes 0 and 1 must be transmitted if the device transmits channel status at all. IEC 958 also indicates that if channel status is not implemented then all bits should be set to zero (including the consumer/professional bit!), in which case a receiver should default to 48 kHz, two-channel, 20 bit audio with no emphasis indicated.

An enhanced mode is also allowed which is basically the standard data plus any additional data such as sample address and source identification.

As far as receivers are concerned there is currently little in the way of insistence in the standards concerning how receivers should behave in the case of certain data combinations, except that the manufacturer should state clearly the data recognized and the actions which will be taken. This situation may be clarified to some extent by the publication of the Engineering Guidelines to the interface, soon to be released by both the AES and the EBU, in which it is expected that a number of equipment classifications will be specified, indicating the level to which the device processes and modifies channel status. It is intended that in the future manufacturers will take more seriously the publication of implementation charts for their devices, much as manufacturers of MIDI-controlled equipment publish such charts in operators' manuals, since this would allow users quickly to identify the roots of any incompatibility.

4.7.5 Overview of channel status in consumer applications

In a way the use of channel status is rather more complicated in consumer applications because of the many types of consumer device and the wide variety of data types that may be transmitted. The format of the basic block is shown in Figure 4.18 and a more detailed breakdown of the first byte (byte 0) is shown in Figure 4.19.

Byte

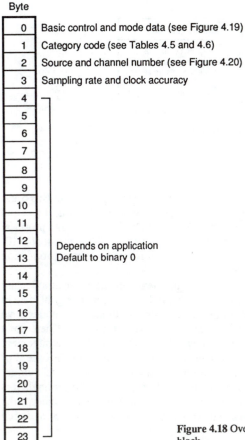

Byte	
0	Basic control and mode data (see Figure 4.19)
1	Category code (see Tables 4.5 and 4.6)
2	Source and channel number (see Figure 4.20)
3	Sampling rate and clock accuracy
4	
5	
6	
7	
8	
9	
10	
11	
12	Depends on application
13	Default to binary 0
14	
15	
16	
17	
18	
19	
20	
21	
22	
23	

Figure 4.18 Overall format of the consumer channel status block

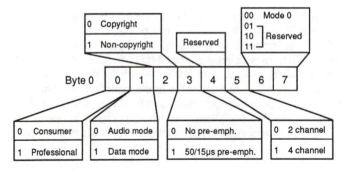

Figure 4.19 Format of byte 0 of consumer channel status

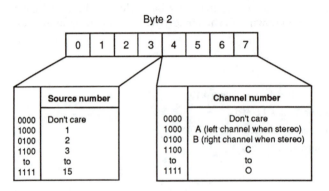

Figure 4.20 Format of byte 2 of consumer channel status

(For a comparison with byte 0 of the professional format see Figure 4.14.) Bits 6 and 7 in the consumer format define the 'mode' of use of the channel status block, and so far only mode 0 is standardized with these bits set to '00', but there is a proposal for a mode 1 to be used for 'software information delivery', with the bits set to '10' respectively, in order to allow the transmission of information about production details on prerecorded media (see below).

The usage of channel status in consumer equipment depends on the mode, and also on the category code which defines the type of device transmitting the data in the second byte (byte 1) of channel status. A large number of category codes have now been defined and they are published in the annexes to IEC 958. There are a number of things to say about them and the way that copy protection management is incorporated with the category code (see section 4.7.7). In mode 0, byte 2 of channel status is used to indicate a source number from 1 to 16 and a channel number (see Figure 4.20), so that in the case of sources with multiple audio channels it is possible to signal which two are being transmitted. Byte 3 is used to indicate the sampling frequency of the source and the clock accuracy of the source (see section 6.4).

4.7.6 Category codes in consumer channel status

The 8 bit category code identifies the source device, allowing subsequent channel status and user data to be interpreted correctly. In the original IEC 958 document only four category codes were defined, and these were as shown in Table 4.5. (Note that category codes are normally written this way round, with the LSB first, which may be confusing since binary numbers are normally written down MSB first.)

At that time it was also proposed that the last bit (bit 15 or the MSB) of the category code should be set to a '1' to indicate 'Music Programme Production Use', in which case bits 32–191 of channel status would be used to carry information relating to the production details of prerecorded programmes, but very soon after the publication of IEC 958 the business of copy protection in DAT systems reached a head and the industry agreed to implement a means of copy management into consumer digital systems, since the record companies were worried that unlimited digital copying would ruin the music business. For this reason, bit 15's function has been changed and is now termed the 'L bit', for reasons which may become clear in the next section, and it is used in the control of copy protection. Most category codes now have two possible values, with the state of the L bit distinguishing between them.

A number of new category code 'families' have arisen since the original IEC 958, the functions of which are defined in annexes, as shown in Table 4.6. (For more detail on subgroupings within these families refer to the standard.)

Table 4.5 Original category codes in IEC 958

Device	Bits 8–15 (LSB to MSB)
General	00000000
CD player	10000000
PCM adaptor	01000000
DAT recorder	11000000

Table 4.6 More recent category code 'families'

Device	Bits 8–15 (LSB to MSB)
Laser optical products	100XXXXL
Signal processors	010XXXXL
Magnetic tape or disk	110XXXXL
Digital audio broadcast	001XXXXL and 0111XXXL
Musical instruments, mics, and other original sources	101XXXXL

4.7.7 SCMS and copy protection

The method of coping with copy management is called SCMS (Serial Copy Management System), and is now implemented on all consumer digital recording equipment. SCMS applies when signals are copied digitally across a consumer format interface, and has no meaning in the professional format. Although SCMS appears not to apply when a signal is copied via an analog interface, in fact there are problems, as will be discussed below.

The principle of SCMS is that a copyright prerecorded signal may be copied only once (provided that the source device is one of the so-called 'white list' of product types from which limited copying is allowed), but no further generations are allowed. This is supposed to allow home users to make a single copy of something they have bought for their own purposes, but to prevent large scale piracy. The copy protection bit which already existed in the consumer format (bit 2) was not sufficient on its own because it did not give any indication of the generation of the copy, so the so-called 'L bit' was brought in, being basically bit 15 of channel status (or the MSB of the category code). The copy protection bit (C bit) in conjunction with the L bit can be used to prevent more than one generation of copying of copyright material. The state of the L bit in effect signifies the 'generation' of the signal, whether '0th generation' (an original prerecorded work) or 1st generation and higher copies, and the C bit's function now signifies whether the source material is copyright or not ('0' = copyright). Unfortunately the complication does not stop here, because although the L bit is normally set so that in the '1' state it represents an original rather than a copy, with laser optical products (such as the CD) and with broadcast receivers it is the other way around!

The upshot of all this is that if a recording device sees that the C bit of a digital source is '0' (copyright material), and the L bit indicates that the source is original prerecorded material, it will allow the copy. If the L bit indicates that the source is already a copy it will disallow the recording. When copyright material is copied from a prerecorded source a flag is recorded on the copy to state 'I am a copy of a copyright source', and when this recording is replayed the L bit will be set to show this, thus disallowing further copies. Extremely thorough coverage of SCMS is to be found in an AES Journal article by Sanchez[24].

4.7.8 SCMS in DAT machines

SCMS was first introduced because of the perceived threat of copying with DAT machines. It works in such machines as follows. There are two DAT category codes: one called simply 'DAT' (code 11000000) and one called 'DAT-P' (code 11000001) – the difference between them is the L bit. On DAT recordings there are also 2 bits in the subcode recorded on the tape that control copy protection and these are called the 'ID6' bits. An SCMS DAT machine looks at the ID6 bits recorded on the tape it is playing to determine how it should set the combination of L and C bits on the digital interface.

If the ID6 on the source tape is 00 (copies allowed) it will set the category to 'DAT' and the C bit to show 'non copyright'. An SCMS machine receiving that signal would also set the recorded ID6 of the copy to 00, since there would be no reason to prevent further copies, and any number of serial digital copies might then be made.

If a DAT recorder sees a 'DAT-P' source it will allow copies no matter what the status of the C bit, whereas if it sees a straightforward 'DAT' source it will only allow a copy if the source is not copy protected. Thus if the ID6 on the source tape is 10 (copies not allowed), the source machine will set the category to 'DAT' and the C bit to 'copyright', and thus the recorder will not be able to copy that tape. If the ID6 on the source tape is 11 (one copy allowed), the machine will set the category to 'DAT-P' and the C bit to 'copyright', and thus a receiver would be able to record the signal (since DAT-P allows copies no matter what the © status), but would know that it was copying © material. The ID6 of the copy would then automatically be set to 10 to prevent further copies.

Concerning copies made between DAT machines and other systems, or between pre-SCMS and SCMS machines, the following applies:

1 *Digital copies from CD*

The CD category code (10000000) when recognized by an SCMS DAT machine will result in a copy whose ID6 is set to 10 (copies not allowed). (The L bit is '0' for original prerecorded material in laser optical products.)

2 *Digital copies from other digital sources*

Copies made from sources having the 'General' category code (00000000) will have their ID6 set to 11, whatever the © status, allowing one further copy only. Sources asserting this code are likely to be such things as A/D convertors and some older DAT machines. It acknowledges that the source of the material is unclear, and might be copyright or might not.

3 *Digital copies to SCMS DAT machines from pre-SCMS machines*

Pre-SCMS machines may use either the 'General' category or the 'DAT' category, depending on when they were made and by whom. They will *not* normally be able to recognize the difference between a recorded ID6 of 11 and an ID6 of 10, because prior to SCMS the machine only had to look at 1 bit to detect © status. Therefore such a machine will normally interpret both codes as indicating that the recording is copy protected (not even allowing one copy), and set the © flag on the digital output.

   Whether or not the receiver will record the data depends on whether the category is 'General' or 'DAT'. If it is 'General' then see 2 above. If it is 'DAT', then not even a single copy will be allowed. The only case in which unlimited copies will be allowed is when the source tape has an ID6 of 00, which is likely to be the case with many tapes recorded on pre-SCMS machines.

4 *Digital copies of recordings made from analog inputs*

Unfortunately, SCMS DAT machines will set the ID6 of analog-sourced recordings to 11, thus allowing only one digital copy. This is a nuisance when the source is a perfectly legitimate non-copyright signal, such as one of your own private recordings.

5 *Digital copies of prerecorded DAT tapes*

The ID6 of prerecorded tapes is set to 11, thus allowing one further copy if using an SCMS machine. If using a pre-SCMS replay machine, the 11 will be interpreted as 10 (see 3 above) and the © bit will be asserted on the interface. A copy will only be possible if the category of the source machine is 'General', but not if it is 'DAT'.

6 *Recording non-copy-protected material on SCMS machines*

There is no way to record completely unprotected material on an SCMS machine, except by feeding it with a digital source having a category code other than 'General' and a recorded ID6 of 00. This might be feasible if you have an early DAT machine. Even material recorded via the SCMS machine's analog inputs cannot be copied beyond a single generation.

7 Digital copying from SCMS machines to pre-SCMS machines

Such copies will only be possible at 48 kHz (or 32 kHz if you have such a tape); 44.1 kHz recordings will be blocked on unmodified machines. Source tapes with ID6 set to either 11 or 10 will cause the CP status to be asserted on the digital interface, and, since pre-SCMS machines tend to ignore the L bit, the copy will not be allowed at all. Source tapes with ID6 set to 00 may be copied.

8 Manual setting of ID6 status

Consumer machines will not allow the ID6 status of tapes to be set, but some recent professional machines will allow this.

4.7.9 Channel status in consumer CD machines

It was originally intended that the first 4 bits of the Q subcode from the CD would be copied into the first 4 bits of the channel status data. The first 4 bits of Q subcode are as follows:

Bit 0   Two or four channel (two channel = '0')
Bit 1   Undefined
Bit 2   Copy protect
Bit 3   Pre-emphasis

Thus apart from bit 0 these are compatible with IEC 958. Since CDs have never implemented a four-channel mode, bit 0 remains in the '0' state which is compatible with the 'consumer' status of bit 0 of IEC 958. It is not clear what would happen if a four-channel CD ever did appear, since it would set bit 0 to '1', signifying 'professional use' of channel status. Other than this, the channel status format of the CD category code is the same as the general format, with the sampling frequency bits set to '0000' to indicate 44.1 kHz. The user bits contain the subcode information, as discussed in section 4.6.2.

4.7.10 Other uses for channel status in consumer formats

MiniDisc and DCC are expected to have the category codes 1001001L and 1100001L respectively. The proposed formats for channel status are so far identical to the general format, although this should be revised in the near future.

The proposed mode for 'software information delivery' or 'music production use' initially intended to use bit 15 of channel status to indicate that such data was contained in bits 32–191 of channel status, but this bit has now been adopted for SCMS purposes. Thus a Japanese proposal exists for Mode 1 of channel status to be set aside for this purpose (signalled in bits 6 and 7). They argue that although the subcode information transmitted in the user bits may often fulfil this purpose it is not independent of the category code, and a format is needed which would be hardware independent. The proposed format would allow for information such as the track titles, technical information relating to playback such as recommended listening levels, and information to control the display of the receiving device. It is a form of menu-based paged system, rather like the teletext system used with television broadcasts.

4.8 The standard multichannel interface (MADI)

Originally proposed in the UK in 1988 by four manufacturers of professional audio equipment (Sony, Neve, Mitsubishi and Solid State Logic), the so-called 'MADI' interface is now an AES and ANSI standard. It was designed to simplify cabling in large installations, especially between multitrack recorders and mixers, and has a lot in common with the format of the two channel interface. The standards concerned are AES10-1991[10] (ANSI S4.43-1991). It is probable that the other organizations which adopted versions of AES3 will eventually adopt AES10 in more or less identical form in the near future. This interface was intentionally designed to be transparent to standard two-channel data, and thus the incorporation of two-channel signals into a MADI multiplex is a relatively straightforward matter. The original channel status, user and aux data remain intact within the multichannel format.

MADI stands for Multichannel Audio Digital Interface; 56 channels of audio are transferred serially in asynchronous form, and consequently the data rate is much higher than that of the two-channel interface. For this reason the data is transmitted either over a coaxial transmission line with 75 ohm termination (not more than 50 m) or over a fibre optic link. The latter is not yet detailed in the standard (although it is mentioned) but is clearly an important consideration for the future since it would allow long distances to be covered. The protocol is based closely on the FDDI (Fibre-Distributed Digital Interface) protocol, and this suggests that fibre optics would be a natural next step[25].

4.8.1 Format of the multichannel interface

The serial data structure is as shown in Figure 4.21, and is divided into subframes which, apart from the preamble area, are identical to AES3 subframes. The preamble is not required here because the interface is synchronized in a different way, and thus the 4 bit slot is replaced with four 'mode bits', the functions of which are labelled in the diagram. Since there are 56 subframes in a frame, bit 0 signifies the start of channel 0 (it goes true for that frame only); bit 1 indicates whether a

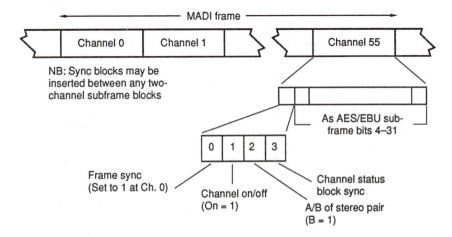

Figure 4.21 Format of the MADI frame

Table 4.7 4/5 bit encoding in MADI

4 bit groups	5 bit codes
0000	11110
0001	01001
0010	10100
0011	10101
0100	01010
0101	01011
0110	01110
0111	01111
1000	10010
1001	10011
1010	10110
1011	10111
1100	11010
1101	11011
1110	11100
1111	11101

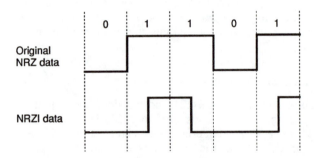

Figure 4.22 An example of the NRZI channel code

particular subframe or audio channel is active (1 for active); bit 2 indicates whether the subframe is either the A or B channel of a two-channel pair derived from an AES3 source (1 for 'B'); and bit 3 indicates the start of a new channel status block for the channel concerned. The audio part of the frame is handled in the same way as in the two-channel interface, and the V, U, C and P bits apply on a per-channel basis, with parity applying over bits 4–31.

The channel code is different from that used in the two-channel version, and another important contrast is that the link transmission rate is independent of the audio sampling frequency or number of channels involved. The highest data transfer rate is at the highest sampling rate (54 kHz) and number of channels (56) times the number of bits per subframe (32), that is $54000 \times 56 \times 32 = 96.768$ Mb/s. It is assumed that the transmitter and receiver will be independently synchronized to the same source of an AES11 reference signal (see Chapter 6), such that they operate at identical sampling frequencies, but the MADI link itself does not synchronize the receiver's sampling clock. The channel-coding process involves two stages: first the 32 bit subframe is divided into groups of 4 bits, and these groups are then encoded into 5 bit words chosen to minimize the DC content of the data signal, according to Table 4.7 (4/5 bit encoding).

The actual *transmission rate* of the data is thus 25% higher than the original data rate, and 32 bit subframes are transmitted as 40 bits. To carry the 4/5-encoded data over the link a '1' is represented by a transition (in either direction) and a '0' by no transition, as shown in the example of Figure 4.22.

Special synchronization symbols are inserted by the transmitter in between encoded subframes, and these take the binary form 11000 10001, transmitted from the left (a pattern which does not arise otherwise). These have the function of synchronizing the receiver (but not its sample clock), and may be inserted between subframes or at the end of the frame in order to fill the total data capacity of the link which is 125 Mb/s ± 100 ppm. The prototype MADI interfaces were designed around AMD's TAXI (Transparent Asynchronous Xmitter/Receiver interface) chips, which are becoming more widely used in high speed computer networks, and these chips normally take care of the insertion of synchronizing symbols so that the transmission rate of the link remains constant.

4.8.2 Electrical characteristics

Since the optical interface is as yet unfinalized the coaxial version will be outlined. It consists of a 75 ohm transmission line terminated in BNC connectors, using coaxial cable with a characteristic impedance of 75 ± 2 ohms and losses of

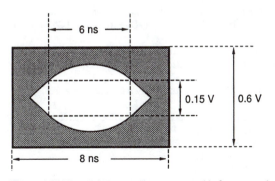

Figure 4.23 The minimum eye pattern acceptable for correct decoding of MADI data

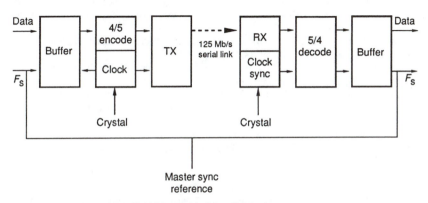

Figure 4.24 Block diagram of MADI transmission and reception

< 0.1 dB/m (1–100 MHz). Suggested driver and receiver circuits are illustrated in the standard, and the receiver is expected to decode data with a minimum eye pattern as shown in Figure 4.23. Equalization is not permitted at the receiver, and distances of up to 50 metres may be covered.

The block diagram of transmitter-to-receiver communication is shown in Figure 4.24. Here the source data is buffered, 4/5 encoded and then formatted with the sync symbols before being transmitted. The receiver extracts the sync symbols, decodes the 4/5 symbols, and a buffer handles any short term variation in timing due to the asynchronous nature of the interface. An external sync reference ensures that the two sampling clocks are locked.

References

1   AES (1985) *AES3-1985 (ANSI S4.40-1985). Serial transmission format for linearly represented digital audio data. Journal of the Audio Engineering Society*, vol. 33, pp. 975–984
2   FINGER, R. (1992) AES3-1992: the revised two channel digital audio interface. *Journal of the Audio Engineering Society*, vol. 40, March, pp. 107–116
3   EBU (1985) *Tech. 3250-E. Specification of the digital audio interface*. Technical Centre of the European Broadcasting Union, Brussels
4   CCIR (1986) *Rec. 647. A digital audio interface for broadcasting studios*. Green Book, vol. 10, pt.1. International Radio Consultative Committee, Geneva
5   CCIR (1990) *Rec. 647 (Mod. F). Draft digital audio interface for broadcast studios*. CCIR, Geneva
6   IEC (1989) *IEC 958. Digital audio interface, first edition*. International Electrotechnical Commission, Geneva
7   EIAJ (1987) *CP-340. A digital audio interface*. Electronic Industries Association of Japan, Tokyo
8   EIAJ (1992) *CP-1201. Digital audio interface (revised)*. Electronic Industries Association of Japan, Tokyo
9   BRITISH STANDARDS INSTITUTE (1989) *BS 7239. Specification for digital audio interface*. British Standards Institute, London
10 AES (1991) *AES10-1991 (ANSI S4.43-1991). Serial multichannel audio digital interface (MADI). Journal of the Audio Engineering Society*, vol. 39, pp. 369–377
11 CCITT (1976, 1980) *Rec. V.11. Electrical characteristics for balanced double-current interchange circuits for general use with integrated circuit equipment in the field of data communications*. International Telegraph and Telephone Consultative Committee
12 EIA. *Industrial electronics bulletin no. 12*. EIA standard RS-422A. Electronics Industries Association, Engineering Dept., Washington, DC
13 DUNN, J. (1991) Considerations for interfacing digital audio equipment to the standards AES3, AES5 and AES11. In *Proceedings of the AES 10th International Conference*, 7–9 September, p. 122, Audio Engineering Society
14 AJEMIAN, R.G. and GRUNDY, A.B. (1990) Fiber optics – the new medium for audio: a tutorial. *Journal of the Audio Engineering Society*, vol. 38, March, pp. 160–175
15 RORDEN, B. and GRAHAM, M. (1992) A proposal for integrating digital audio distribution into TV production. *JSMPTE*, September, pp. 606–608
16 GILCHRIST, N. (1989) Coordination signals in the professional digital audio interface. In *Proceedings of the AES/EBU Interface Conference*, 12–13 September, pp. 13–15, Audio Engineering Society British Section
17 KOMLY, A. and VIALLEVIEILLE, A. (1989) Programme labelling in the user channel. In *Proceedings of the AES/EBU Interface Conference*, 12–13 September, pp. 28–51, Audio Engineering Society British Section
18 ISO 3309 (1984) *Information processing systems – data communications – high level data link frame structure*. International Organization for Standardization
19 AES18 (1992) Format for the user data channel of the AES digital audio interface. *Journal of the Audio Engineering Society*, vol. 40, no. 3, March, pp. 167–183

20 NUNN, J. P. (1992) Ancillary data in the AES/EBU digital audio interface. In *Proceedings of the 1st NAB Radio Montreux Symposium*, 10–13 June, pp. 29–41

21 Report of SC-2-4 signal labelling and ancillary data working group of SC-2 subcommittee on digital audio. *Journal of the Audio Engineering Society*, vol. 40, no. 3, March 1992, pp. 184–185

22 DAT CONFERENCE PART V (1986) *Digital audio taperecorder system (RDAT). Recommended design standard.*

23 PHILIPS (1992) *ITTS Interactive Text Transmission System*. Reference document, July

24 SANCHEZ (1994) An understanding and implementation of the SCMS serial copy management system for digital audio transmission. *Journal of the Audio Engineering Society*, vol. 42, no. 3, March, pp. 162–186

25 WILTON, P. (1989) *MADI (Multichannel audio digital interface)*. In *Proceedings of the AES/EBU Interface Conference*, 12–13 September, pp. 117–130, Audio Engineering Society British Section

Chapter 5

Digital audio interfaces – II

Interfaces other than the standard two-channel and multichannel will be described and discussed in this chapter. For example, there are a number of interfaces which have been introduced by specific manufacturers and which are normally only found on that manufacturer's products. Additionally we have chosen to include brief coverage of NICAM dual sound in syncs (DSIS), since this is a system widely used in Europe for the digital interconnection of studio centres in broadcasting. We have also chosen to mention the MIDI standard (not a real-time audio interface) and ISDN, since although these are not digital audio interfaces as such they are both used to carry audio information at times. Not included in the book is the whole topic of carrying audio over computer networks, such as Ethernet and FDDI (except MADI), the key reason being that this technology is only just emerging in professional audio applications and is not in the same 'family' as the real-time digital audio interfaces described here. Neither have we included SCSI (the Small Computer Systems Interface), since this is a computer interface for the parallel interconnection of peripherals such as disk drives, although there are commercial devices available which will convert AES/EBU digital audio data into a format which can be read into a computer via an SCSI interface.

5.1 The Sony digital interface (SDIF)

The most common Sony interface is SDIF-2, and is intended for the transfer of one channel of digital audio information per physical link, at a resolution of up to 20 bits (although most devices only make use of 16). It is an interface which has also been used on equipment other than Sony's, for the sake of compatibility, but it is likely

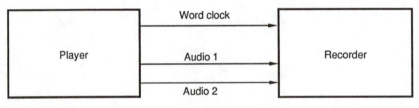

Figure 5.1 Sony SDIF-2 interconnection for two audio channels

Table 5.1 Pinouts for differential SDIF multichannel interface

Pin	Function
1, 2	Ch. 1 (–/+)
3, 4	Ch. 2 (–/+)
5, 6	Ch. 3 (–/+)
etc	etc
to	to
47, 48	Ch. 24 (–/+)
49, 50	NC

that this standard will gradually decline as the standard two-channel interface (see section 4.2) becomes more widely used.

The interface, as used on most two-channel equipment, is unbalanced and uses 75 ohm coaxial cable terminating in 75 ohm BNC-type connectors, one for each audio channel. TTL-compatible electrical levels (0–5 V) are used. The audio data is accompanied by a word clock signal on a separate physical link (see Figure 5.1), which is a square wave at the sampling frequency used to synchronize the receiver's sample clock. Sony's multitrack machines use SDIF also, but with a differential electrical interface conforming to RS-422 standards (see section 1.6.2) and using 50 pin D-type multiway connectors, the pinouts of which are shown in Table 5.1. A single BNC connector carries word clock as before.

In each audio sample period, the equivalent of 32 bits of data is transmitted over each physical link, although only the first 29 bits of the word are considered valid, since the last three bit-cell periods are divided into two cells of one-and-a-half times the normal duration, violating the NRZ code in order to act as a synchronizing pattern. As shown in Figure 5.2, 20 bits of audio data are transmitted with the MSB first (although typically only 16 bits are used), followed by nine control or user bits (although the user bits are rarely employed). A block structure is created for the control/user bits which repeats once every 256 sample periods, signalled using the block sync flag in bit 29 of the first word of the block. The resulting data rate is 1.53 Mb/s at 48 kHz sampling rate and 1.21 Mb/s at 44.1 kHz.

The SDIF-2 interface is used mainly for the transfer of audio data from Sony professional digital audio equipment, particularly the PCM-1610 and 1630, but also from semi-professional Sony equipment which has been modified to give digital inputs and outputs (such as the PCM-701 and various DAT machines). It is also used on a number of disk-based workstations for the loading and unloading of audio data. It is not recommended for use over long distances and it is important that the coaxial leads for each channel and the word clock are kept to the same length otherwise timing errors may arise. Problems occasionally arise with third-party implementations of this interface that do not use the 1.5 bit-cell sync pattern at the end of words, requiring some trial and error involving delays of the data signal with relation to the separate word clock in order for the link to function correctly.

5.2 Mitsubishi digital interfaces

Mitsubishi's ProDigi format tape machines use a digital interface similar to SDIF, but not compatible with it. Separate electrical interconnections are used for each audio channel. Interfaces labelled 'Dub A' and 'Dub B' are 16 channel interfaces

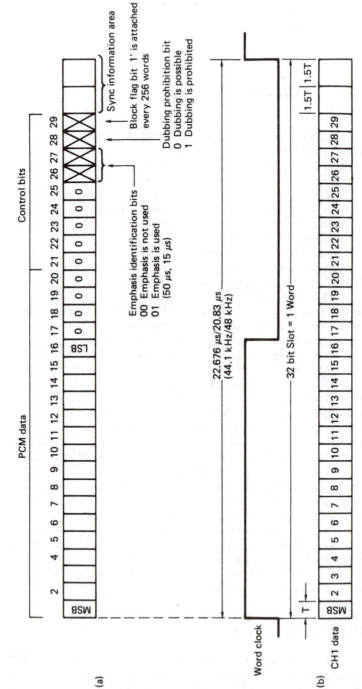

(a)

PCM data

Control bits

MSB   LSB   0 0 0 0 0 0 0 0 0 0

2 4 5 6 7 8 9 10 11 12 13 14 15 16 17 18 19 20 21 22 23 24 25 26 27 28 29

Sync information area

Block flag bit '1' is attached every 256 words

Dubbing prohibition bit
0 Dubbing is possible
1 Dubbing is prohibited

Emphasis identification bits
00 Emphasis is not used
01 Emphasis is used
(50 µs, 15 µs)

22.676 µs/20.83 µs
(44.1 kHz/48 kHz)

32 bit Slot = 1 Word

1.5T | 1.5T

Word clock

T

(b)

CH1 data

MSB

2 3 4 5 6 7 8 9 10 11 12 13 14 15 16 17 18 19 20 21 22 23 24 25 26 27 28 29

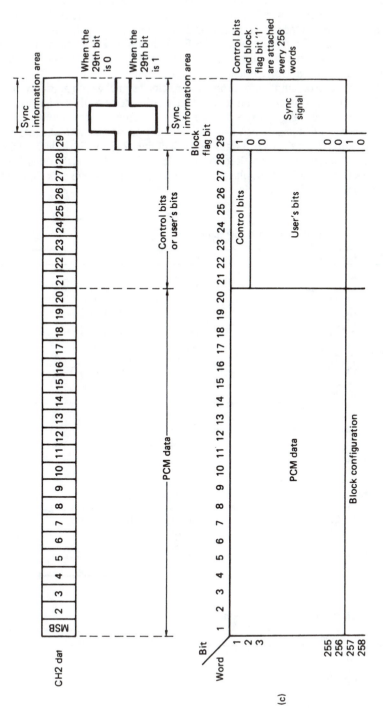

Figure 5.2 At (a) is the block content of the SDIF signal; (b) shows the synchronizing patterns used for data reception. At (c) user bits form a block which is synchronized every 256 sample periods

found on multitrack machines, handling respectively tracks 1–16 and 17–32. These interfaces terminate in 50 way D-type connectors and utilize differential balanced drivers and receivers. One sample period is divided into 32 bit cells, only the first 16 of which are used for sample data (MSB first), the rest being set to zero. There is no sync pattern within the audio data (such as there is between bits 30–32 in the SDIF-2 format). The audio data is accompanied by a separate bit clock (1.536 MHz square wave at 48 kHz sampling rate) and a word clock signal which goes low only for the first bit cell of each 32 bit audio data word (unlike SDIF which uses a sampling rate square wave), as shown in Figure 5.3. Status information is passed over two separate channels, which take the same format as an audio channel but carry information about the record status of each of the 32 channels of the ProDigi tape machine. One status channel (Rec 'A') handles tracks 1–16, and the other (Rec 'B') handles tracks 17–32 of a multitrack machine. The pin assignments for these connectors are shown in Table 5.2.

Table 5.2(a) Pinouts for Mitsubishi 'Dub A' connector

Pin	Function	Pin	Function
1, 18	Ch. 1 (+/–)	10, 27	Ch. 10 (+/–)
2, 19	Ch. 2 (+/–)	11, 28	Ch. 11 (+/–)
3, 20	Ch. 3 (+/–)	12, 29	Ch. 12 (+/–)
4, 21	Ch. 4 (+/–)	13, 30	Ch. 13 (+/–)
5, 22	Ch. 5 (+/–)	14, 31	Ch. 14 (+/–)
6, 23	Ch. 6 (+/–)	15, 32	Ch. 15 (+/–)
7, 24	Ch. 7 (+/–)	16, 33	Ch. 16 (+/–)
8, 25	Ch. 8 (+/–)	17, 50	GND
9, 26	Ch. 9 (+/–)	34, 35	Bit clock (+/–)
36, 37	WCLK (+/–)	38, 39	Rec A (+/–)
40, 41	Rec B (+/–)		

Table 5.2(b) Pinouts for Mitsubishi 'Dub B' connector

Pin	Function	Pin	Function
1, 18	Ch. 17 (+/–)	9, 26	Ch. 25 (+/–)
2, 19	Ch. 18 (+/–)	10, 27	Ch. 26 (+/–)
3, 20	Ch. 19 (+/–)	11, 28	Ch. 27 (+/–)
4, 21	Ch. 20 (+/–)	12, 29	Ch. 28 (+/–)
5, 22	Ch. 21 (+/–)	13, 30	Ch. 29 (+/–)
6, 23	Ch. 22 (+/–)	14, 31	Ch. 30 (+/–)
7, 24	Ch. 23 (+/–)	15, 32	Ch. 31 (+/–)
8, 25	Ch. 24 (+/–)	16, 33	Ch. 32 (+/–)
17, 50	GND		

Table 5.3 Pinouts for Mitsubishi 'Dub C' connectors

Pin	Function
1, 14	Left (+/–)
2, 15	Right (+/–)
5, 18	Bit clock (+/–)
6, 19	WCLK (+/–)
7, 20	Master clock (+/–)
12, 25	GND

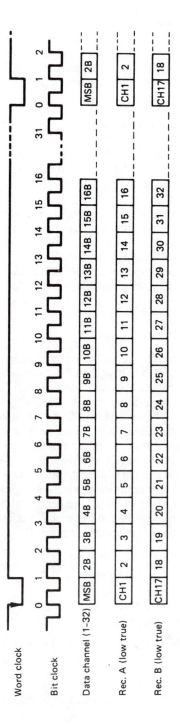

Figure 5.3 Data format of the Mitsubishi multitrack interface

Mitsubishi interfaces labelled 'Dub C' are two-channel interfaces. These terminate in 25 way D-type connectors and utilize unbalanced drivers and receivers. One sample period is divided into 24 bit cells, only the first 16 or 20 of which are normally used, depending on the resolution of the recording in question. Again audio data is accompanied by a separate bit clock (1.152 MHz square wave at 48 kHz sampling rate) and a word clock signal taking the same form as the multitrack version. No record status information is carried over this interface, but an additional 'master clock' is offered at 2.304 MHz. The pin assignments are shown in Table 5.3.

5.3 Sony to Mitsubishi conversion

Comparing the SDIF format with the Mitsubishi multichannel format it may be appreciated that to interconnect the two would only require minor modifications to the signals. Both have a 32 bit structure, MSB first, with only 16 bits being used for audio resolution, the difference being the sync pattern in Sony's bits 29–32, plus the fact that the Mitsubishi does not include control and user bits. The word clock of the Sony is a square wave at the sampling frequency, whereas that of Mitsubishi only goes low for one bit period, but simple logic could convert from one to the other. If transferring from Sony to Mitsubishi it would be necessary also to derive a bit clock, and this could be multiplied up from the word clock using a suitable phase-locked loop. Commercial interfaces are available which perform this task neatly.

5.4 Yamaha interface

Yamaha digital audio equipment is often equipped with a 'Cascade' connector to allow for a number of devices to be operated in cascade, such that the two-channel mix outputs of a mixer, for example, may be fed into a further mixer to be combined with another mixed group of channels.

This interface terminates in an eight-pin DIN-type connector, as shown in Figure 5.4, and carries two channels of 24 bit audio data over an RS-422-standard differential line. The two channels of data are multiplexed over a single serial link, with a 32 bit word of left channel data followed by a 32 bit word of right channel data (the 24 bits of audio are sent LSB first, followed by eight zeros). Word clock alternates between low state for the left channel and high state for the right channel, as shown in Figure 5.5. Coils of 20 µH are connected between pins 6 and 7 and ground to enable suppression of radio frequency interference. The OUT socket is only enabled when its pin 8 is connected to ground.

Pin-outs

1 WCLK +
2 GND
3 Audio data –
4 WCLK –
5 Audio data +
6 20 µH coil to GND
7 20 µH coil to GND
8 GND (in), ENABLE (out)

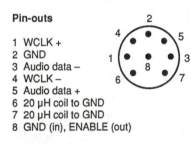

Figure 5.4 Pinouts of the Yamaha 'cascade' interface

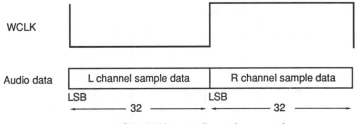

Only 24 bits actually used per sample

Figure 5.5 Data format of the Yamaha 'cascade' interface

5.5 Ampex parallel audio interface

Ampex devised a four-channel digital audio interface for the D2 video format which used the same electro-mechanical interface as the parallel digital video interface (see Chapter 8). This is a balanced parallel format, terminating in 25 pin D-type

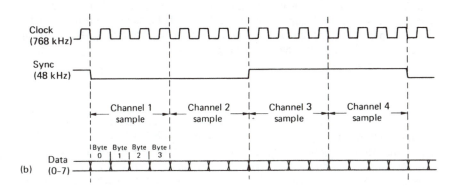

Contact	Signal line	Contact	Signal line
1	Clock		
2	System ground	14	Clock ret.
3	Data 7 (MSB)	15	System ground
4	Data 6	16	Data 7 ret.
5	Data 5	17	Data 6 ret.
6	Data 4	18	Data 5 ret.
7	Data 3	19	Data 4 ret.
8	Data 2	20	Data 3 ret.
9	Data 1	21	Data 2 ret.
10	Data 0	22	Data 1 ret.
11	Spare	23	Data 0 ret.
12	Sync	24	Spare ret.
13	Chassis shield	25	Sync ret.

Figure 5.6 In the Ampex four-channel parallel digital audio interconnect, 4 byte channel samples are sent serially in one sample period. The same connections are used as for the video interconnect except that an additional signal, SYNC, shown at (a) is needed to specify which samples correspond to which channel as in (b)

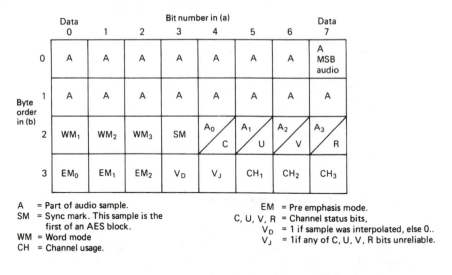

	Data 0	1	2	Bit number in (a) 3	4	5	6	Data 7
0	A	A	A	A	A	A	A	A MSB audio
1	A	A	A	A	A	A	A	A
2	WM$_1$	WM$_2$	WM$_3$	SM	A$_0$ / C	A$_1$ / U	A$_2$ / V	A$_3$ / R
3	EM$_0$	EM$_1$	EM$_2$	V$_D$	V$_J$	CH$_1$	CH$_2$	CH$_3$

Byte order in (b)

A   = Part of audio sample.
SM   = Sync mark. This sample is the
       first of an AES block.
WM   = Word mode
CH   = Channel usage.

EM = Pre emphasis mode.
C, U, V, R = Channel status bits,
V$_D$ = 1 if sample was interpolated, else 0..
V$_J$ = 1if any of C, U, V, R bits unreliable.

Figure 5.6(c) The 4 bytes corresponding to each audio channel are assembled as shown here

connectors, in which 8 bits of data are accompanied by a clock and sync signal, using the pinouts shown in Figure 5.6(a). Samples for four channels of audio data are transmitted within one sample period, retaining commonality of data with the standard two-channel AES format (see section 4.2), by transmitting 32 bits for each channel in four lots of 8 bits (see Figure 5.6(b)). The clock signal, which has transitions in the middle of a time slot, is at 16 times the sampling frequency (768 kHz), and marks the 16 time slots involved. The 32 bits of data received for each channel are formatted as shown in Figure 5.6(c), and it will be seen that there are 16 audio bits, with four optional LSBs that share a position with the normal VUCP data of the AES subframe. Some of the important parts of AES channel status have been extracted and given slots of their own, namely 'word mode' (derived from byte 2, bits 0–2, indicating use of auxiliary bits and length of audio sample), 'emphasis mode' (which copies the emphasis bits of channel status), and 'channel usage' (which indicates the channel mode, from byte 1, bits 1–3). The 'SM' bit signifies the start of a new channel status block.

5.6 MIDI

MIDI is not strictly a digital audio interface – it is a remote control interface for musical instruments and related studio equipment – but it has a mode in which digital audio information may be carried in non-real time between devices, normally used for transferring sample data between music samplers, or between a sampler and a computer for editing purposes. This mode will be covered here for the sake of completeness, but for a more detailed coverage of MIDI control the reader is referred to Rumsey[1].

MIDI (the Musical Instrument Digital Interface) is a unidirectional serial interface which transfers 8 bit data messages asynchronously at 31.25 kbaud. The

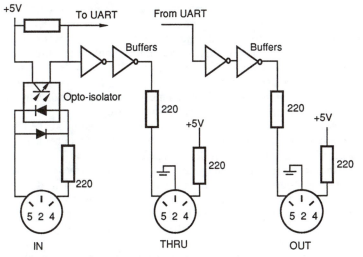

Figure 5.7 Hardware configuration of the MIDI interface

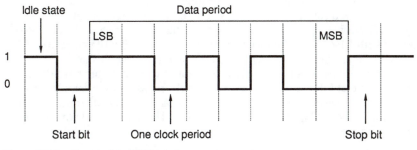

Figure 5.8 Data format of the MIDI interface

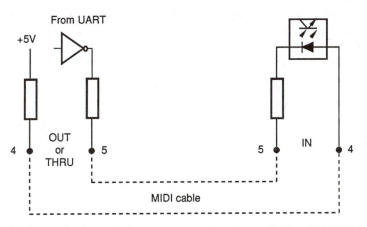

Figure 5.9 A current loop is formed between a MIDI OUT or THRU and a MIDI IN

electrical interface is shown in Figure 5.7, and the format of a single message byte is shown in Figure 5.8. A current loop is formed between an OUT or THRU of a transmitter and an IN of the receiver, when connected with a shielded twisted pair cable terminated in a standard five-pin DIN connector using only the centre three pins, as shown in Figure 5.9. Typically a MIDI message consists of a 'status byte' followed by a number of 'data bytes', the status byte beginning (MSB) with a binary '1' and indicating the type of message and the channel number, and the data bytes beginning with a binary '0', containing the data of the message.

The group of messages containing digital audio sample information is called the 'universal non-real-time' messages. The standard allows for transfer either using a single MIDI link from sampler to receiving device (open-loop), or with links in both directions so that the receiving device can signal correct reception to the transmitter (closed-loop). The closed-loop system is a more secure way of transferring data, as the receiver can request re-transmission of data if any is erroneous or lost for whatever reason. In a closed loop, the MIDI OUT of the sampler would be connected to the MIDI IN of the receiver, and the OUT of the receiver (not the THRU) would go back to the IN of the sampler.

Two types of message can be used in this protocol: principally those which transmit and signal the attributes of sample data itself, and those which are for 'housekeeping' purposes (known as handshaking). Handshaking data is returned from the receiver to the transmitter in order to signal such things as whether the data has been received correctly or not. All sample dump messages are in the so-called 'System Exclusive' format; thus they begin with the status byte hexadecimal (&) F0 and end with &F7. &F0 is followed by the Universal Non Real Time header, &7E, which, in conjunction with relevant sub IDs, signifies a sample dump.

Assuming a closed loop, the sequence of events for a sample dump would follow the protocol that follows. Firstly, the receiver should ask for the sample it wants. (This message may also be initiated from a front-panel control on the transmitter in the case of an open loop, rather than over MIDI.) The message is Dump Request, and takes the form:

&[F0] [7E] [nn] [03] [ss] [ss] [F7]

where [nn] is the channel number, and [ss] [ss] is the number of the requested sample as stored in the sampler itself (LSByte first). Thus up to 2^{14} samples could be selected by the two [ss] bytes (MSB of each must be a zero for data). The receiver should then check that the number of the requested sample is within the range of its stored samples, and ignore the message if it is not. If the sample number is legal, the transmission should begin, preceded by the Dump Header. This describes all the vital statistics of the sample to be transmitted, so that the receiving device knows exactly what form it takes, how many bits there are per sample, what the sampling rate is, and so on. The header takes the form:

&[F0] [7E] [nn] [01] [ss] [ss] [ee] [ff] [ff] [ff] [gg] [gg] [gg] [hh] [hh] [hh] [ii] [ii] [ii] [jj] [F7]

where:

[nn]	channel number
[ss] [ss]	stored sample number (LSByte first)
[ee]	number of bits per sample (from 8 to 28)
[ff] [ff] [ff]	sample period in nanoseconds (i.e. 1/sample rate; LSByte first)

[gg] [gg] [gg]   sample length (number of words of 8–28 bits)
[hh] [hh] [hh]   sustain loop start (in number of words from sample start; LSByte
                  first)
[ii] [ii] [ii]   sustain loop end (as above)
[jj]   loop type (00 = forward, 01=forward/backward)

Within 2 seconds, the receiver should send one of two handshaking messages back to the transmitter: either ACK (acknowledge) if it is capable of accepting the data described by the header, or CANCEL if it is not. ACK takes the form:

&[F0] [7E] [nn] [7F] [pp] [F7]

and CANCEL takes the form:

&[F0] [7E] [nn] [7D] [pp] [F7]

where [pp] is the packet number (see below). In the case of this first message to acknowledge the header, the packet number is unnecessary, or could be set to zero.

On receiving ACK, the transmitter should start sending packets of sample data. CANCEL should abort the exchange. If nothing is received by the transmitter within the 2 second period, it will assume that an open loop is in use and begin to send packets of data anyway.

A third acknowledgement is possible after receipt of the header: this is WAIT, which signals that the transmitter should wait until another message is received. The receiver could then send one of the other two messages when it became ready to operate. Therefore it is recommended that WAIT be sent initially, rather than nothing, if a delay in responding greater than 2 seconds is envisaged.

Sample data is transmitted in 120 byte packets, regardless of the size of the sample or the number of bits per sample. This has been judged to be the minimum number of bytes that could be stored in a modern MIDI device's input buffer if none were processed before the packet ended, when added to the status bytes and packet header. It thus avoids buffer overflow. A packet takes the following form:

&[F0] [7E] [nn] [02] [kk] <120 bytes> [LL] [F7]

where [kk] is a running count of packets, from 0 to 127, starting again from zero if necessary. [LL] is a checksum, which is the XOR of the preceding bytes of the packet, namely. :

[7E] [nn] [02] [kk] <120 bytes>

to detect errors.

The 120 bytes of MIDI sample data represent stored sample words in 2, 3, or 4 byte groups as follows (remember that each byte can only use 7 bits, because MSB = 0). For samplers using between 8 and 14 bits per sample, 2 bytes represent a sample word (60 words per packet); for samplers using between 15 and 21 bits per sample, 3 bytes represent a sample word (40 words per packet); and for samplers using between 22 and 28 bits per sample, 4 bytes represent a sample word (30 words per packet).

Data should be left-justified within the groups of MIDI bytes if the sample data does not completely fill the space available, and unused bits should be made zero. Checksums of each packet are to be verified by the receiver, and the return link used to send either ACK if it is correct (see above), or NAK (not acknowledge) if it is not. NAK takes the form:

&[F0] [7E] [nn] [7E] [pp] [F7]

where [pp] is the packet number. NAK should result in a re-transmission of the offending packet.

It should be noted that not all devices may have the ability to receive packets out of sequence, although the presence of individual packet numbers makes it possible to transmit any single packet to a device capable of handling this. There is possible room for confusion if the number of packets exceeds 128, as the sequence of packet numbers will begin again from zero.

It may also be appreciated that although the universality of this format makes for compatibility, it may also be slower than an individual manufacturer's dump format, due to the need to accommodate and signal a wide variety of sample types. Nonetheless, some commercial samplers are provided with faster interfaces, capable of transferring data at two or four times normal speed. It follows that any receiver would also be required to operate at this higher rate.

5.7 NICAM Dual Sound in Syncs (DSIS)

Within television networks it is important these days to be able to carry stereo sound over long distances with an accompanying vision signal. In parts of Europe (especially the UK) it is now increasingly common for time-compressed stereo digital audio data to be inserted into the line sync period of the television waveform, using a form of data reduction called NICAM 728[2]. NICAM 728 is the data reduction scheme and coding method chosen by the UK and parts of Scandinavia for terrestrial stereo TV broadcasts, and is recommended by the EBU for any European organization intending to broadcast a new stereo service. Dual sound in syncs (DSIS)[3] is used for carrying NICAM-coded audio data with a vision signal over microwave links between studio centres, and from studio centres to transmitters. It has the advantage that the vision and stereo audio signals are carried together

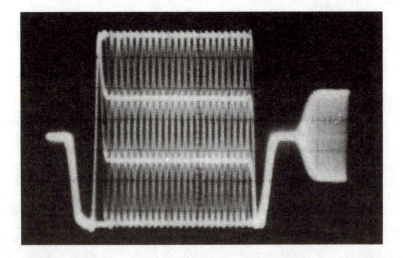

Figure 5.10 Eye pattern of a typical NICAM DSIS signal in the video line sync period

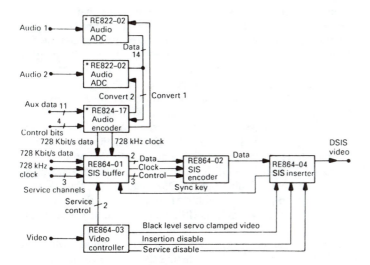

Figure 5.11(a) Schematic diagram of a commercial DSIS encoder (Courtesy of RE Instruments)

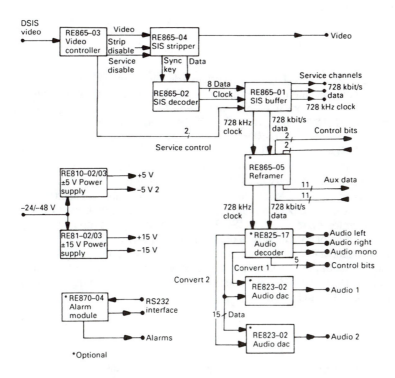

Figure 5.11(b) Schematic diagram of a commercial DSIS decoder (Courtesy of RE Instruments)

over a single link, both reducing the costs involved and ensuring that the correct signals remain together. DSIS is not an audio interface typically found on standard items of equipment, but a number of broadcast studio centres use it in conjunction with AES-style digital audio routing at the studio centre, as a means of keeping the signal in an all-digital form from source to consumer. Transcoders exist which convert data from a standard AES two-channel format to DSIS and vice versa. A useful overview of a successful installation of this kind may be found in Evans[4].

In NICAM DSIS the NICAM-coded audio data is split into bit pairs and modulated onto the TV waveform using a four-level coding method in which each voltage level corresponds to one of the four states of a pair of bits. Figure 5.10 shows an enlarged photograph of the eye pattern of digital audio information in the line sync period. TV signals which include DSIS information cannot be transmitted, processed or monitored without stripping out the DSIS audio data; thus DSIS is normally only used for point-to-point links within a network, rather than within the studio centre. A block diagram of a typical DSIS coder and decoder is shown in Figure 5.11.

A number of data reduction processes have been introduced recently which are considerably more efficient than NICAM at reducing the bit rate without unduly affecting sound quality, such as the so-called 'ISO Layers'[5], but these were not available at the time when the digital stereo TV service was being put into place in the UK and Scandinavia.

5.8 ISDN

ISDN, the Integrated Services Digital Network, is a development in the telecommunications field which is gaining wider usage. ISDN is an extension of the digital telephone network to the consumer, providing two 64 kb/s digital channels which can be connected to ISDN terminals anywhere in the world simply by dialling. Data of virtually any kind may be transferred over the dialled-up link, and potential applications for ISDN include audio transfer. A comprehensive description of ISDN can be found in a book by John Griffiths of British Telecom[6].

Since the total usable capacity of an ISDN 'B' connection is only 128 kb/s it is not possible to carry linear PCM data at normal audio resolutions over such a link, but it is possible to carry moderately high quality stereo digital audio at this rate using a data reduction system such as ISO Layer 3[5], or to achieve higher rates by combining more than one ISDN link to obtain data rates of, say, 256 or 384 kb/s[7]. Using the higher data rates one could adopt ISO Layer 1 or 2 coding, or alternatively one could adopt one of the non-ISO commercial low bit rate coding systems such as apt-X 100[8] or Dolby AC-2[9]. Such coders will normally accept a digital audio input in the standard two-channel form.

References

1   RUMSEY, F.J. (1994) *MIDI Systems and Control*, 2nd edn., Focal Press, Oxford
2   BBC (1988) NICAM 728; specification for two additional digital sound channels with System I television. *BBC Engineering Information*, August
3   HOLDER, J.E. *et al.* (1984) A two-channel sound-in-syncs transmission system. In *Proceedings of the International Broadcast Convention*, pp. 345–348, Institution of Electrical Engineers, London

4   EVANS, P. (1990) Digital audio in the broadcast centre. *EBU Technical Review*, no. 241/242, June–August, European Broadcasting Union

5   BRANDENBURG, K. and STOLL, G. (1992) The ISO/MPEG audio codec: a generic standard for coding of high quality digital audio. Presented at the *92nd AES Convention, Vienna, Austria*, 24–27 March, preprint 3336

6   GRIFFITHS, J. (1991) *ISDN explained*

7   BURKHARDTSMAIER, B. *et al.* (1992) The ISDN MusicTAXI. Presented at the *92nd AES Convention, Vienna, Austria*, 24–27 March, preprint 3344

8   SMYTH, M. and SMYTH, S. (1991) APT-X 100: a low delay, low bit rate, sub-band ADPCM audio coder for broadcasting. In *Proceedings of the AES 10th International Conference*, 7–9 September, pp. 41–56, Audio Engineering Society

9   FIELDER, L.D. (1991) AC-2: a family of low complexity transform-based music coders. In *Proceedings of the AES 10th International Conference*, 7–9 September, pp. 57–70, Audio Engineering Society

Chapter 6

Synchronization in digital audio interfacing

6.1 The need for synchronization

Unlike analog audio, digital audio has a discrete-time structure, because it is a sampled signal in which the samples may be further grouped into frames and blocks having a certain time duration. If digital audio devices are to communicate with each other, or if digital signals are to be combined in any way, then it is important that they are synchronized to a common reference in order that the sampling frequencies of the devices are identical and do not drift with relation to each other. It is not enough for two devices to be running at *nominally* the same sampling frequency (say, both at 44.1 kHz), since between the sampling clocks of professional audio equipment it is possible for differences in frequency of up to ±10 parts per million (ppm) to exist, and even a very slow drift means that two devices are not truly synchronous.

The audible effect resulting from a non-synchronous signal drifting with relation to a sync reference or another signal is usually the occurrence of a glitch or click at the difference frequency between the signal and the reference, typically at an audio level around 50 dB below the signal, due to the repetition or dropping of samples. This will appear when attempting to mix two digital audio signals whose sampling rates differ by a small amount, or when attempting to decode a signal such as an unlocked consumer source by a professional system which is locked to a fixed reference. This said, it is not always easy to detect asynchronous operation by listening, even though sample slippage is occurring. Some systems may not operate at all if presented with asynchronous signals.

Furthermore, when digital audio is used with analog or digital video, the sampling rate of the audio needs to be locked to the video reference signal, and also to any timecode signals which may be used. In single studio operations the problem of ensuring lock to a common clock is not as great as it is in a multi-studio centre, or where digital audio signals arrive from remote locations. In such cases either the remote signals must be synchronized to the local sample clock as they arrive, or the remote studio must somehow be fed with the same reference signal as the local studio. A number of approaches may be used to ensure that this happens and they will be explained in this chapter.

Another topic related to synchronization will also be examined here, and that is the importance of short term clock stability in digitally interfaced audio systems. It is important to distinguish between clock stability requirements in interfacing and clock stability in convertors, although the two are related to some extent.

6.2 Choice of sync reference

6.2.1 AES recommendations

There are now AES recommendations for the synchronization of digital audio signals, documented in AES11-1991[1]. They state that preferably all machines should be able to lock to a reference signal, which should take the form of a standard two-channel interface signal (see section 4.2) whose sampling frequency is stable within a certain tolerance, and that all machines should have a separate input for such a synchronizing signal. If this procedure is not adopted then it is possible for a device to lock to the clock embedded in the channel code of the AES-format audio input signal – a technique known as 'genlock' synchronization.

In the AES11 standard signals are considered synchronous when they have identical sample rates, but phase errors are allowed to exist between the reference clock and received/transmitted digital signals in order to allow for effects such as cable propagation delays, phase-locked loop errors and other electrical effects. Input signal frame edges must lie within ±25% of the reference signal's frame edge (taken as the leading edge of the 'X' preamble), and output signals within ±5% (see Figure 6.1), although tighter accuracy than this is preferable because otherwise an unacceptable build up of delay may arise when equipment is cascaded. A phase error of ±25% of the frame period is actually quite considerable, corresponding to a timing difference of around 5 μs at 48 kHz, and thus the specification should be readily achievable in most circumstances. For example, a cable length of 58 metres typically gives rise to a delay of only 1 bit (0.32 μs @ 48 kHz) in the interface signal. If a number of frames' delay exists between the audio input and output of a device, this delay should be stated.

The AES11 reference signal may either contain programme or not. If it does not contain programme it may be digital silence or simply the sync preamble with the rest of the frame inactive. Two grades of reference are specified: Grade 1, having a long term frequency accuracy of ±1 ppm (part per million), and Grade 2, having a long term accuracy of ±10 ppm. The Grade 2 signal conforms to the standard AES sample frequency recommendation for digital audio equipment (AES5-1984[2]), and

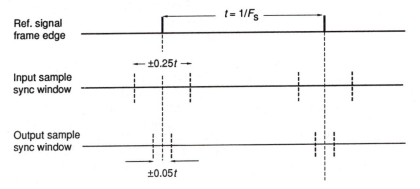

Ref. signal
frame edge

$t = 1/F_s$

±0.25t

Input sample
sync window

Output sample
sync window

±0.05t

Figure 6.1 Timing limits recommended for synchronous signals in the AES11 standard

is intended for use within a single studio which has no immediate technical reason for greater accuracy, whereas Grade 1 is a tighter specification and is intended for the synchronization of complete studio centres (as well as single studios if required). Byte 4, bits 0 and 1 of the channel status data of the reference signal (see section 4.7) indicate the grade of reference in use (00=default, 01=Grade 1, 10=Grade 2, 11=reserved). It is also specified in this standard that the capture range of oscillators in devices designed to lock to external synchronizing signals should be ± 2 ppm for Grade 1 and ± 50 ppm for Grade 2. (Grade 1 reference equipment is only expected to lock to other Grade 1 references.)

These AES recommendations, of which this is only a summary, were only defined in 1991 and equipment manufacturers have so far been slow to adopt the provision of a separate AES-format input for a synchronizing signal. It is expected that this will become more common over the next few years.

6.2.2 Other forms of external sync reference

Currently, digital audio recorders are provided with a wide range of sync inputs, and most systems may be operated in the external or internal sync modes. In the internal sync mode a system is locked to its own crystal oscillator, which in professional equipment should be accurate within ± 10 parts per million (ppm) if it conforms to AES recommended practice (AES5-1984), but which in consumer equipment may be much less accurate than this (see section 6.4).

In the external sync mode the system should lock to one of its sync inputs, which may either be selectable using a switch, or be selected automatically based on an order of priority depending on the mode of operation (for this the user should refer to the operations manual of the device concerned). Typical sync inputs are word clock (WCLK), which is normally a square-wave TTL-level signal (0–5 V) at the sampling rate, usually available on a BNC-type connector and common on equipment using the Sony interface (SDIF); and 'composite video', which is a video reference signal consisting of either normal picture information or just 'black and burst' (a video signal with a blacked-out picture). In all cases one machine or source must be considered to be the 'master', supplying the sync reference to the whole system, and the others as 'slaves'.

WCLK may be 'daisy-chained' (looped through) between devices in cases where the AES/EBU interface is not available. A 'sync' light is usually provided on the front panel of a device (or under a cover) to indicate good lock to an external or internal clock, and this may flash or go out if the system cannot lock to the clock concerned, perhaps because it has too much jitter, is at too low a level, conflicts with another clock, or because it is not at a sampling rate which can be accepted by the system.

Video sync (composite sync) is often used when a system is operated within a video environment, and where a digital recorder is to be referenced to the same sync reference as the video machines in a system. This is useful when synchronizing audio and video transports during recording and replay, using either a synchronizer or a video editor, when timecode is only used initially during the pre-roll, whereafter machines are released to their own video sync reference. It also allows for timecode to be recorded synchronously on the audio machine, since the timecode generator used to stripe the tape can be locked to video syncs also. The relationship between video frame rates and audio sampling frequencies is discussed further in section 6.6.1.

6.3 Distribution of sync references

It is appropriate to consider digital audio as similar to video when approaching the subject of sync distribution, especially in large systems. Consequently, it is advisable to use a central high quality sync signal generator (the equivalent of a video sync pulse generator, or SPG), the output of which is made available widely around the studio centre, using digital distribution amplifiers (DDAs) to supply different outlets in a 'star' configuration. In video operations this is often called 'house sync', and such a term may become used in digital audio as well. Each digital device in the system may then be connected to this house sync signal, and each should be set to operate in the external sync mode. Once digital audio equipment provides a separate AES-format sync input on all equipment it will be advisable for the 'house sync' signal to be either a Grade 1 or Grade 2 AES11 reference signal (see above). Since long cable runs or poor quality interfaces can distort digital signals, resulting in timing jitter (see section 6.4.3), it may be advisable to install local reference signal generators slaved to the house sync master generator as a means of providing a 'clean' local reference within a studio. Until such time as AES11 reference inputs become available on audio products one might use word clock or a video reference signal instead.

Using the technique of central sync-signal distribution (see Figure 6.2) it becomes possible to treat all devices as slaves to the sync generator, rather than each one locking to the audio output of the previous one. In this case there is no 'master' machine, since the sync generator acts as the 'master'. It requires that all machines in the system operate at the same sampling rate, unless a sample rate convertor or signal synchronizer is used (see section 6.5).

Alternatively, in a small studio, it may be uneconomical or impractical to use a separate SPG, and in such cases one device in the studio must be designated as the master. This device would then effectively act as the SPG, operating in the internal

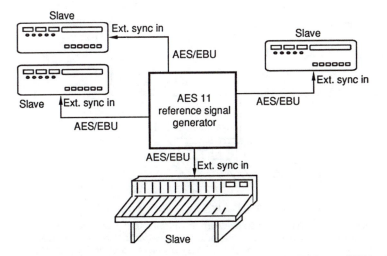

Figure 6.2 All devices within a studio may be synchronized by distributing an AES11 reference signal to external AES sync inputs (if present)

sync mode, with all other devices operating in the external sync mode and slaving
to it (see Figure 6.3). If a digital mixer were to be used then it could be used as the
SPG, but alternatively it would be possible to use a tape recorder, disk system or
other device with a stable clock. In such a configuration it would be necessary either
to use AES/EBU interfaces for all interconnection (in which case it would be

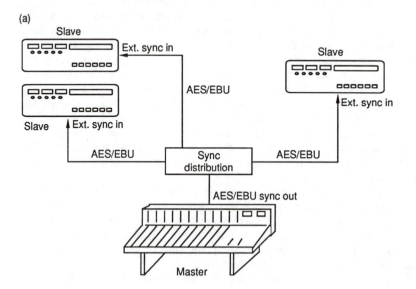

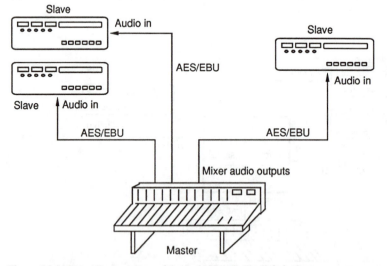

Figure 6.3 (a) In small systems one of the devices, such as a digital mixer, may act as a sync
reference for the others; alternatively, as shown in (b), devices may lock to their AES-format audio
inputs, although they may only do so when recording. For synchronous replay, devices normally
require a separate external sync input, unless they have been modified to lock to the digital audio
input on replay as well

possible to operate in the 'genlock'-type mode described above, where all devices derive a clock from their digital audio inputs) or to distribute a separate word clock or AES sync signal from the master. In the genlock configuration, the danger exists of timing errors being compounded as delays introduced by one device are passed serially down the signal chain.

In situations where sources are widely spread apart, perhaps even being fed in from remote sites (as might be the case in broadcast operations) the distribution of a single sync reference to all devices becomes very difficult or impossible. In such cases it may be necessary to re-time external 'wild' feeds to the house sync before they can be connected to in-house equipment, and for this purpose a sample rate synchronizer should be used (see section 6.5).

6.4 Clock accuracy considerations

As mentioned above, there are a number of different considerations concerning clock accuracy and the importance of stability which apply at different points in the signal chain. There is the accuracy of the audio sample clock used for A/D and D/A conversion, which will have a direct effect on sound quality, and there is the accuracy of the external reference signal. There is also the question of timing stability in digital audio signals which have travelled over interconnects such as those described in this book, which may have suffered distortions of various kinds. Since the audio sample clock used for conversion in externally synchronized systems must be locked in some way either to the digital audio input clock or the sync reference, it is common for instabilities in either of these latter signals to affect the stability of the audio sample clock (although they need not), and this depends on how the clock is extracted from the digital input signal and the nature of the timing error. Furthermore, clock instability resulting from distortion and interference in the digital interface makes the signal more difficult to decode.

6.4.1 Audio sampling frequency

It is important to separate the discussion of the long term sample frequency accuracy from that in the short term. Short term inaccuracy is called 'jitter', and long term inaccuracy would manifest itself as drift, or wow and flutter in extreme cases. Jitter will be covered in the next section.

For professional equipment, the nominal sampling frequency should be accurate to within ±10 ppm if it conforms to AES5 recommendations (although the standard only explicitly states this at 48 kHz). This corresponds to an allowable peak drift in the sample period of ±0.21 ns at a sampling frequency of 48 kHz, but implies nothing about the *rate* at which that modulation takes place. When a device 'free runs' it is locked to its own internal oscillator, which for fixed sampling frequencies is normally a crystal oscillator capable of high accuracy. However, in variable-speed modes a crystal oscillator cannot easily be used and thus some form of voltage-controlled or other oscillator may take its place, having a less stable frequency, and in such cases a device is not expected to meet the stability requirements of AES5 or AES11.

In consumer equipment the sampling frequency is normally less carefully specified and controlled, often making it difficult to interconnect consumer and professional equipment without the use of a sample rate convertor or a synchronizer.

Table 6.1 Consumer clock accuracy in channel status of IEC 958

Bits 24–27	Sampling frequency
0 0 0 0	44.1 kHz
0 1 0 0	48 kHz
1 1 0 0	32 kHz

Bits 28–29	Clock accuracy
0 0	Level II
0 1	Level III
1 0	Level I
1 1	Reserved

IEC 958 specifies three levels of sampling frequency accuracy: Level I ('high' accuracy) = ±50 ppm; Level II (normal accuracy) = ±1000 ppm; and Level III (variable pitch shifted clock mode), which is undefined except to say that the frequency range is likely to be ±12.5% of the nominal sampling frequency. Again nothing is said about the rate of sample clock modulation. Consumer sampling frequency and clock accuracy are indicated in bits 24–27 and 28–29 of channel status in the digital audio interface signal (see section 4.7), according to Table 6.1. There is no such indication in professional channel status, except in AES11-type reference signals (see above).

At the professional limit of ±10 ppm, a nominal sample clock of 48 kHz could range over the limits 47 999.52 Hz to 48 000.48 Hz – a speed tolerance of 0.001% – whereas a normal accuracy consumer device at the same nominal sampling frequency could be anything from 47 956 Hz to 48 048 Hz – a speed tolerance of ±0.1%.

6.4.2 Sample clock jitter and effects on sound quality

Short term timing irregularities in sample clocks may affect sound quality in devices such as A/D and D/A convertors, and sample rate convertors. This is due to the modulation in the time domain of the sample instant (see section 2.7.4), resulting in low level signal products within the audio spectrum. The important features of jitter are its peak amplitude and its rate, since the effect on sound quality is dependent on both of these factors taken together. Shelton[3], by calculating the rms signal-to-noise ratio resulting from random jitter, shows that timing irregularities as low as 5 ns may be significant for 16 bit digital audio systems over a range of signal frequencies, and that the criteria are even more stringent at higher resolutions and at high frequencies. The effects are summarized in Figure 6.4.

When jitter is periodic rather than random, it results in the equivalent of 'flutter', and the effect when applied to the sample clock in the conversion of a sinusoidal audio signal is to produce sidebands on either side of the original audio signal due to phase modulation, whose spacing is equal to the jitter frequency. Julian Dunn[4] has shown that the level of the jitter sideband (R_j) with relation to the signal is given by:

$$R_j \text{ (dB)} = 20 \log (J\omega_i/4)$$

where J is the peak-to-peak amplitude of the jitter and ω_i is the audio signal

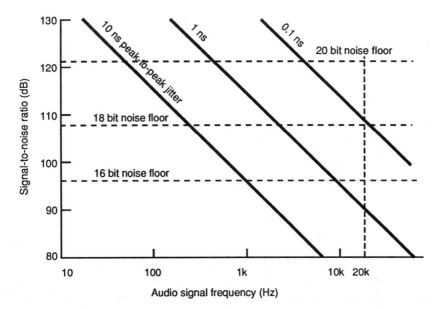

Figure 6.4 Effects of sample clock jitter on signal-to-noise ratio at different frequencies, compared with theoretical noise floors of systems with different resolutions (after W. T. Shelton, with permission)

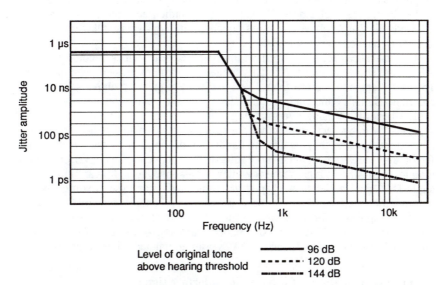

Figure 6.5 Sample clock jitter amplitude at different frequencies required for just audible modulation noise on a worst-case audio signal (after J. Dunn, with permission)

frequency. Using this formula he shows that for sinusoidal jitter with an amplitude of 500 ps, a maximum level 20 kHz audio signal will produce sidebands at −96.1 dB relative to the amplitude of the tone.

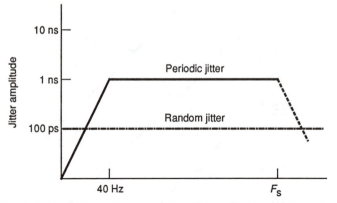

Figure 6.6 AES11 recommendations for sample clock jitter in 16 bit systems

What is important, though, is the *audibility* of jitter-induced products, and Dunn[4,5] has attempted to calculate this based on an analysis of the resulting spectrum using accepted audibility curves, based on critical band masking theory, assuming that the audio signal is replayed at a high listening level (120 dB SPL). As shown in Figure 6.5, which plots jitter amplitude against jitter frequency (not audio frequency) for just-audible modulation noise on a worst-case audio signal, the jitter amplitude may in fact be very high (>1 µs) at low jitter frequencies (up to around 250 Hz) since the sidebands will be masked at all audio frequencies, but the amount allowed falls sharply above this jitter frequency, although it may still be up to ±10 ns at jitter frequencies up to 400 Hz.

The tolerances for jitter on the sampling frequency clock recommended in AES11 are shown in Figure 6.6. This suggests maximum peak-to-peak jitter for a 16 bit system of no more than ±1 ns at jitter frequencies above 40 Hz. AES recommendations concerning sample clock jitter are currently under revision.

The benefits of using oversampling convertors are not entirely clear. Story[6] suggests that they reject sample clock jitter to a greater extent than non-oversampling convertors, because they reduce the bandwidth of the jitter by the oversampling factor, thus making them much more suitable than conventional convertors. Dunn, though, suggests that high frequency jitter (well above the audio band) in a delta-sigma DAC may modulate the shaped ultrasonic modulator noise, creating products within critical parts of the audio spectrum. A comprehensive study by Chris Dunn and Malcolm Hawksford[7] attempts to survey the effects of interface-induced jitter on different types of DAC, and this paper warrants close study by those whose business it is to design high quality DACs with digital audio interfaces.

6.4.3 Causes and effects of jitter on the interface signal

Timing irregularities may arise in signals transferred over a digital audio interface due to a number of factors. These may include bandwidth limitations of the interconnect, and the effects of induced noise and other signals. Furthermore, the transmitted signal may already have some jitter, either because the source did not properly reject incoming jitter from an input signal, or because its own free-running clock was unstable. AES3 specifies that data transitions on the interface should occur within ±20 ns of an ideal jitter-free clock, but the jitter on the interface signal must be distinguished from sample clock jitter (although since sample clocks may be derived from interface signals the two are clearly related). In real products it is normal for interface transition jitter to be well below 20 ns, but when devices which pass on input jitter to their digital outputs are cascaded it is possible for specifications to exceed this value after a number of stages. The degree to which a device passes on jitter is known as its jitter transfer function.

The received signal from a standard two-channel AES interface will have an eye pattern which depends on amplitude and timing irregularities. Amplitude errors will close the eye vertically and timing errors will close it horizontally, and the limits for correct decoding are laid out in the specification of the interface. Nonetheless, some receivers are better than others at decoding data with a poor eye pattern and this has partly to do with the frequency response of the phase-locked loop in the receiver and its lock-in range (see below). It also depends on the part of the signal from which the decoder extracts its clock, since some transitions are decidedly more unstable than others when the link is poor (see below). Decoders which are very tolerant of poor input signals may at the same time be bad at rejecting jitter; therefore, although they may decode the signal, the resulting sound quality may be poor if the signal is converted within the device without further rejection of jitter, and the device may pass on excessive jitter at its output. There are examples of devices which will not decode correctly even data which has jitter specifications within the AES3 limits.

Dunn[4] and Dunn and Hawksford[7] have both carried out simulations of the effects of link bandwidth reduction on standard two-channel interface signals. The important conclusions of their work are as follows. When a link suffers high frequency loss there will be a reduction in amplitude of the shorter pulses and a slowing in rise and fall times at transitions, the effect of which is to delay the zero-crossing transition after short pulses less than the delay after longer pulses. This variable delay is effectively jitter, and is solely a result of high frequency loss. Figure 6.7 shows the HF loss model which was used in both studies (which, although simplistic, is considered a good starting point for analysis), the time constant of which is RC, and Figure 6.8 shows a comparison between simulated bi-phase mark data at time constants of 200 ns and 50 ns, showing clearly that at 200 ns the shorter pulses are more attenuated than the longer (a time constant of 200 ns corresponds to a roll-off of −3dB at 0.8 MHz, whereas 50 ns corresponds to a similar roll-off at around 3.18 MHz). Dunn's results show clearly that for links with a bandwidth of less than around 3 MHz the jitter suffered by transitions in the main part of the AES subframe (the audio data time slots) is far greater than that suffered by the penultimate transition of the Y preamble, as shown in Figure 6.9. Dunn and Hawksford also provide convincing evidence that the jitter is highly correlated with the audio signal, and is affected by the difference between the number of zeros and ones in the signal.

The implication of this is that it is greatly preferable to derive a stable sample clock from one of the reliable preamble transitions than from the audio data slot transitions, although there is evidence that many devices do use the data transitions. One interface receiver chip adjusts its PLL on every negative-going transition in the interface signal, for example. Since transitions in the audio data part of the subframe are not only more sensitive to line-induced jitter, but also determined by the audio signal, it is even possible for a signal on the B audio channel to modulate the sampling clock such that jitter sidebands appear in the A channel that are tonally related to the signal in the B channel.

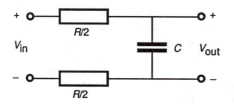

Figure 6.7 High frequency loss model used in simulating the effects of cables

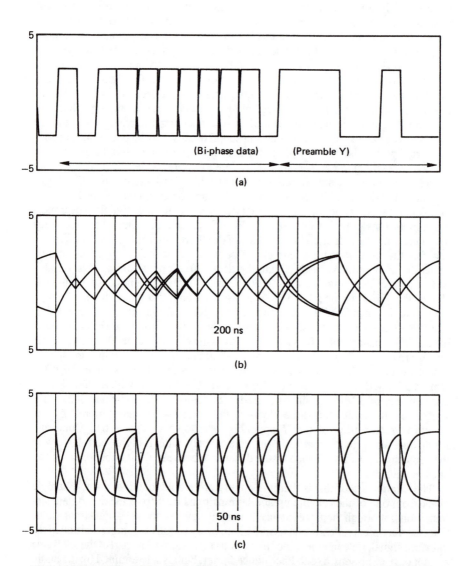

Figure 6.8 A simulation of the effects of high frequency loss on an AES3-format signal. At (a) the original signal is shown (both polarities are superimposed for clarity), whilst (b) and (c) show the data waveform at link time constants of 200 ns and 50 ns respectively. (Reproduced from J. Dunn, with permission)

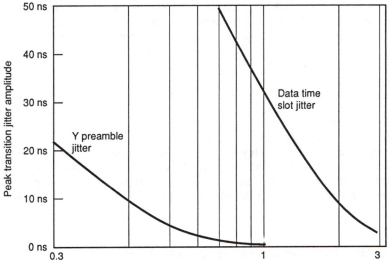

Figure 6.9 Effect of link bandwidth on preamble and data time slot jitter, showing how the Y preamble is affected much less by reduced bandwidth than the data edges. (Reproduced from J. Dunn, with permission)

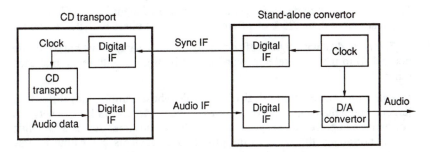

Figure 6.10 In one high quality CD player with a separate convertor, rather than deriving the convertor clock from the incoming audio data, the convertor clock is fed back to the transport via a separate interface in order to maintain synchronization of the incoming data

The rejection of jitter by the receiver depends principally on the frequency response of its PLL, and in general the narrower the response of the PLL and the lower its cut-off frequency the better the rejection of jitter. The problem with this is that such PLLs may not lock up as quickly as wide bandwidth PLLs, and may not lock over a particularly wide frequency range. A solution suggested by Dunn and Hawksford is to use a wideband PLL in series with a low bandwidth version which switches in after conditions have stabilized. Alternatively RAM buffering may be used between the interface decoder and the convertor, the data being clocked out of the buffer to the convertor under control of a more stable clock source. In a hi-fi system it is possible that the stable clock source's frequency could be independent of the incoming data clock, provided that the buffer was of sufficient size to accommodate the maximum timing error between the two over the duration of, say, a CD, but the better solution adopted by some manufacturers in two-box CD players is to generate a synchronizing clock in the convertor which is fed back to reference the speed of the CD transport and thus the rate of data coming over the digital interface (see Figure 6.10).

In a professional digital audio system where all devices in the system are to be locked to a common sampling frequency clock it would normally be necessary to lock any convertors to an external reference. An AES11-style reference signal derived from a central generator might have suffered similar degradations to an audio signal travelling between two devices and thus exhibit transition timing jitter; therefore in such cases, especially in areas where high quality D/A conversion is required, it would be advisable either to re-clock the reference signal or to use a local high quality reference generator slaved to the central AES11 generator, with which to reference the convertors.

6.5 Use and function of sample rate synchronizers

A sample rate synchronizer may be used to lock a digital audio signal to a reference signal. It could also perform the useful function of re-clocking distorted remote signals to remove short term timing errors (jitter). Three main approaches to sample rate synchronization are necessary, depending on the operational requirement, and each will be discussed in turn. These are:

(a) *Frame alignment*: to deal with signals which are of identical sample rate but which are more than ±25% of a frame period out of phase with the reference.

(b) *Buffering*: to deal with signals of nominally the same sample rate but which are not locked to the same reference and thus drift slightly with relation to each other.

(c) *Sample rate conversion*: to deal with signals whose sample rates differ by a larger amount than implied in (b), such as between 44.1 and 48 kHz, or between consumer and professional systems in which the consumer device's sampling rate is nominally the same as the professional device's but within a large tolerance (see section 6.4, above).

6.5.1 Frame alignment

It may be necessary to correct for timing errors in signals synchronous to the master clock, but which have travelled long distances. These will have been delayed and thus be out of phase with the reference. This is more properly referred to as 'frame alignment' and is only necessary when a signal is more than ±25% of a frame period delayed with reference to the sync reference. Propagation delays are not great, however: for example, an AES/EBU signal must travel some 3.7 km down a typical cable before it is delayed by one sample period; thus it is most unlikely that such a situation will arise in real operational environments unless a large static phase error has been introduced in the sample clock of an incoming signal due to it having been cascaded through a number of devices operating in the genlock mode (see section 6.3).

In order to conform to AES recommendations, frame alignment should rephase the signal to bring it within ±5% of a frame period compared with the sync reference. Input signals less than ±25% adrift are also expected to be brought within this ±5% limit at the output, but this is normally performed within the device itself. Reframing of signals more than ±25% adrift may be performed within the device, but if not an external reframer would be required.

6.5.2 Buffering

For signals of nominally the same frequency but very slightly adrift it is possible to use a simple buffer store synchronizer, such as those described by Gilchrist[8] and also by Parker[9]. In this type of synchronizer, a typical block diagram of which is pictured in Figure 6.11, audio samples are written sequentially into successive addresses of a solid state memory configured in the FIFO manner. These samples are read out of the memory a short time later, clocked by the reference signal, the buffer providing a short term store to accommodate the variation in input and output rates. If the output rate is slightly faster than the input rate then the buffer will gradually become empty, and if it is slower than the input rate the buffer will gradually become full, requiring action at some point to avoid losing data or repeating samples because the buffer cannot be infinitely large. At such a time the read address is reset to the mid point of the buffer, resulting in a slight discontinuity in the audio signal. This discontinuity may be arranged to occur within silent passages of the programme, or alternatively a short crossfade may be introduced at the reset point to 'hide' the discontinuity.

Buffer store synchronizers have the advantage that most of the time the audio signal is copied bit for bit between input and output, with discontinuities only occurring once every so many minutes, depending on the size of the buffer and the discrepancy in input and output sampling rates. The larger the buffer the longer the gaps between buffer resets, or the greater the discrepancy between sampling rates which may be accommodated. The price for using a larger buffer is a longer delay between input and output, and this must be chosen with the operational requirement in mind. Using a buffer store capable of holding 480 samples, for example, a delay of around 5 ms would result, and buffer resets would occur every 8.3 minutes if the sample rates were at the extremes of the AES5 tolerance of ±10 ppm.

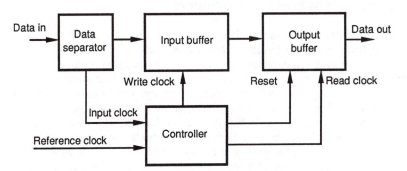

Figure 6.11 Block diagram of an example of a simple buffer store synchronizer

6.5.3 Sample rate conversion

For signals whose sample rates differ by too great an amount to be handled by a buffer store synchronizer it will be necessary to employ sample rate conversion. This can be used to convert digital interface signals from one rate to another (say from 44.1 to 48 kHz) without passing through the analog domain. Sample rate conversion is not truly a transparent process but modern convertors introduce minimal side effects.

The most basic form of sample rate conversion involves the translation of samples at one fixed rate to a new fixed rate, related by a simple fractional ratio. Fractional-ratio conversion involves the mathematical interpolation of samples at the new rate based on the values of samples at the old rate. Digital filtering is used to calculate the amplitudes of the new samples such that they are mathematically correct based on the impulse response of original samples, after lowpass filtering with an upper limit of the Nyquist frequency of the original sampling rate. A clock rate common to both sample rates is used to control the interpolation process. Using this method, some output samples will coincide with input samples, but only a limited number of possibilities exist for the interval between input and output samples. Such a process is nominally jitter free.

If the input and output sampling rates have a variable or non-simple relationship, output samples may be required at any interval in between input samples. This requires an interpolator with many more clock phases than for fractional-ratio conversion, the intention being to pick a clock phase which most closely corresponds with the desired output sample instant at which to calculate the necessary coefficient. There will clearly be a timing error, the audible result of which is equivalent to the effect of jitter (see section 6.4.2), and this may be made smaller by increasing the number of possible interpolator phases. If the input sampling rate is continuously varied (as it might be in variable-speed searching or cueing) the position of interpolated samples with relation to original samples must vary also, and this requires real-time calculation of filter phase.

Errors in sample rate conversion should be designed so as to result in noise modulation below the noise floor of a 16 bit system, and preferably lower. For example, one such convertor is quoted as introducing distortion and noise at −105 dB ref. 1 kHz at 0 dB FS (equivalent to 18 bit noise performance).

6.6 Considerations in video environments

Audio and video/film have traditionally required synchronization for the purposes of achieving lip sync. Film and video are both discrete in that they have frames, and when audio was analog synchronizing to sufficient accuracy for lip sync caused no difficulty. Now that audio is also digital, it too is discrete information and more accurate synchronizing with video becomes necessary. In environments where digital audio is used with video signals, it is important for the audio sampling rate to be locked to the same master clock as the video reference. The same applies to timecode signals which may be used with digital audio and video equipment. A number of proposals exist for incorporating timecode within the standard two-channel interface, each of which has different merits, and these will be discussed below.

6.6.1 Relationships between video frame rates and audio sampling rates

People using the PAL or SECAM television systems are fortunate in that there is a simple integer relationship between the sampling rate of 48 kHz used in digital audio systems for TV and the video frame rate of 25 Hz (there are 1920 samples per frame). There is also a simple relationship between the other standard sampling rates of 44.1 and 32 kHz and the PAL/SECAM frame rate, as shown in Table 6.2. Users of NTSC TV systems (such as the USA and Japan) are less fortunate because the TV frame rate is 30/1.001 (roughly 29.97) frames per second, resulting in a non-integer relationship with standard audio sampling rates. The sampling rate of 44.056 kHz was introduced in digital audio recording systems which used NTSC VTRs, since this resulted in an integer relationship with the frame rate, but this sampling rate is used infrequently today.

The standard two-channel interface's channel status block structure repeats at 192 sample intervals, and thus in 48 kHz systems there are exactly ten audio interface frames per PAL/SECAM video frame, simplifying the synchronization of information contained in channel status with junctures in the TV signal, and making possible the carrying of EBU timecode signals in channel status as described below.

As described by Shelton[10] and others[11,12] it is desirable to source a master audio reference signal centrally within a studio operation, just as a video reference is centrally sourced, and these two references are normally locked to the same highly stable rubidium reference, which in turn may be locked to a standard reference frequency broadcast by a national transmitter. As shown in Figure 6.12, the overall master clock contained in a central apparatus room will be used to lock the video SPG (distributed as 'black and burst' video), the audio reference signal generator (distributed as a standard AES11 reference signal) and the colour subcarrier synthesizer. The video reference is used in turn to lock timecode generators. Appropriate master clock frequencies and division ratios may be devised for the application in question.

There is currently no universal agreement on the appropriate sync point between audio and video reference signals, although proposals have been made for the use of the leading edge of line sync of both line 1 and line 6 of the video frame in PAL systems. The situation is complicated with NTSC video since there are not an integer number of audio samples per frame, requiring that a certain phase relationship between audio and video repeats itself every so many frames, with the sync point being specified at midnight.

An alternative approach to the locking of the sampling rate of digital audio equipment to a video reference is found on some equipment, especially digital tape recorders. Here the audio device accepts a video sync reference and derives its sample clock by appropriate multiplication and division of this reference internally. DIP switches may be provided on the audio device to select the appropriate frame rate, so that the correct ratio results.

Table 6.2 Audio samples per TV frame in PAL and NTSC systems

Sampling rate	PAL/SECAM	NTSC
32 kHz	1280	16 016/15
44.1 kHz	1764	147 147/100
48 kHz	1920	8008/5

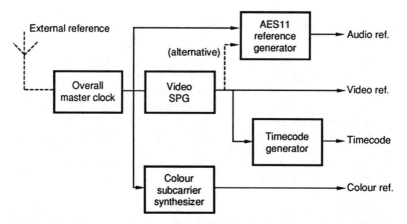

Figure 6.12 In video environments all audio, video and timecode sync references should be locked to a common clock

6.6.2 Referencing of VTRs with digital audio tracks

Video tape recorders (VTRs), both analog and digital, are often equipped with digital audio tracks. Digital VTRs are really data recorders with audio and video interfaces. The great majority of the data is video samples and the head rotation is locked to the video field timing. Part of each track is reserved for audio data, which uses the same heads and much of the same circuitry as the video on a time-division basis. As a result, the audio sampling rate has to be locked to video timing so that the correct number of audio and video samples can be assembled in order to record a track.

At the moment the only audio sampling rate supported by digital VTRs is 48 kHz. This causes no difficulty in 625/50 systems as there are exactly 960 sample periods in a field and a phase-locked loop can easily multiply the vertical rate to produce a video synchronous 48 kHz clock. Alternatively, line rate can be multiplied by 384/125. However, the 0.1% offset of 525 line systems makes the actual field rate 59.94 Hz. The fields are slightly longer than at 60 Hz and in 60 fields there will be exactly 48 048 sample periods. Unfortunately this number does not divide by 60 without a remainder. The smallest number of fields which contain a whole number of samples is five. This makes the generation of the audio sampling clock more difficult, but it can be obtained by multiplying line rate by 1144/375.

When a DVTR is recording using analog audio inputs, the heads rotate locked to input video, which also determines the audio sampling rate for the ADCs in the machine and there is no difficulty. On replay, the synchronism between video and audio is locked in the recording and the audio sampling rate on the output will be locked to station reference. Any receiving device will need to slave to the replay audio sampling rate. On recording with a digital audio input, it is necessary that the digital audio source is slaved to the input video of the DVTR. This can be achieved either by taking a video-derived audio sampling rate reference from the DVTR to the source, or by using a source which can derive its own sampling rate from a video input. The same video timing is then routed to both the source and the DVTR.

If these steps are not followed, there could be an audio sampling rate mismatch between the source and destination, and inevitably periodic corruption will occur.

With modern crystal-controlled devices it is surprising how long an unlocked system can run between corruptions, and it is easy mistakenly to think a system is locked simply because it works for a short while when in fact it is not and will shows signs of distress if monitored for longer.

In some VTRs with digital audio tracks it has been possible for there to arise a phase relationship between the digital audio outputs of a VTR and the video output which is different each time the VTR is turned on, causing difficulties in the digital transfer of audio and video data to another VTR. When the phase relationship is such that the incoming digital audio signal to a VTR lies right on the boundary between two sample slots of its own reference there is often the possibility of sample slips or repeats when timing jitter causes vacillation of sync between two sample periods. This highlights the importance of fixing the audio/video phase relationship.

Manufacturers of video recorders, though, claim that the solution is not as simple as it seems, since in editing systems the VTR may switch between different video sync references depending on the operational mode. Should the audio phase always follow the video reference selection of the VTR? In this case audio reframing would be required at regular points in the signal chain. A likely interim solution will be that at least the phase of the house audio reference signal with relation to the house video sync will be specified, providing a known reference sync point for any reframing which might be required in combined audio/video systems.

6.6.3 Timecode in the standard two-channel interface

In section 4.7 the possibility of including 'sample address' codes in the channel status data of the interface was introduced. Such a code is capable of representing a count of the number of samples elapsed since midnight in a binary form, and when there is a simple relationship between the audio sample rate and the video frame rate as there is with PAL TV signals, it is relatively straightforward to convert sample address codes into the equivalent value in hours, minutes, seconds and frames used in SMPTE/EBU timecode.

At a sample rate of 48 kHz the sample address is updated once every 4 ms, which is ten times per video frame. At NTSC frame rates the transcoding is less easy, especially since NTSC video requires 'drop-frame' SMPTE timecode to accommodate the non-integer number of frames per second. A further potential difficulty is that the rate of update and method of transcoding for sample address codes is dependent upon the audio sampling frequency, but since this is normally fixed within one studio centre the issue is not particularly important.

Discussions have been running for a number of years on an appropriate method for encoding SMPTE/EBU timecodes in a suitable form within the channel status or user bits of the audio interface. At the time of writing there is no published standard for this purpose, although a number of European broadcasters have adopted the non-standard procedure described below. There is also potential for incorporating timecode messages into the HDLC packet scheme determined for the user bit channel in AES18 (see section 4.6.1). The advantage of this option is that the data transfer rate in AES18 is independent of the audio sample rate, within limits, and thus the timecode could possibly be asynchronous with the audio, video, or both, if necessary. Such a scheme makes it difficult to derive a phase reference between the video frame edge and the timecode frame edge, but additional bits are proposed in this option to indicate the phase offset between the two at regular points in the video frame[12].

An important proposal exists, already adopted by a number of European broadcasters[13], which replaces the local sample address code in bytes 14–17 of channel status with 4 bytes of conventional SMPTE/EBU timecode. These bytes contain the BCD (Binary Coded Decimal) time values for hours, minutes, seconds and frames, as replayed from the device in question, in the following manner:

Byte 14 = frames (a 4 bit BCD value for both tens and units of frames)
Byte 15 = seconds (ditto for seconds)
Byte 16 = minutes (ditto for minutes)
Byte 17 = hours (ditto for hours)

The time-of-day sample address bytes (18–21) are replaced by time-of-day timecode in the same way. At 48 kHz with EBU timecode the timecode value is thus repeated ten times per frame.

Thames TV has unilaterally adopted bit 3, byte 22 of channel status to indicate that this type of timecode exists in place of sample address codes, and it is possible that a similar approach may become part of an AES standard in the future, but it should be noted that in NTSC video systems using drop-frame timecode the method would be less suitable, as it would also be in systems where the timecode was asynchronous with the audio. This scheme is cheaper than transcoding sample addresses to timecode or vice versa at points in the chain, provided that tape recorders and other systems insert the timecode at source and receivers are capable of extracting it. Thames TV quotes two manufacturers which already provide these facilities.

It has also been suggested that the format used for inserting timecode into the subcode area of the professional DAT system might be adapted for use in the digital audio interface. In DAT recorders, SMPTE/EBU timecode is transcoded to DAT timecode internally. The DAT timecode represents hours, minutes, seconds and *DAT frames* (at 33.33 fps), together with an 11 bit value to indicate the difference in phase between the start of the DAT frame and the start of the related SMPTE/EBU frame. It also uses 3 bits to identify the frame rate of the original SMPTE/EBU timecode.

Although it is not appropriate for the digital audio interface timecode signal to be tied to the DAT frame structure, an analogous concept has been suggested by Rumsey[14] which could be used to code timecode within channel status – with an hours, minutes, seconds and frames BCD code in place of the sample address, together with 2 bits to indicate the SMPTE/EBU frame rate, and a separate byte to indicate the phase offset of the first audio block in a new timecode frame from the video frame edge, measured in audio samples.

Since channel status blocks are shorter than video frames, the same timecode value would be repeated for a number of blocks, depending on the sample and frame rates. This would use more space than sample address code provides, although since only 28 bits are actually needed to represent a BCD timecode frame (tens of hours and tens of frames only need 2 bits), the remaining 4 bits of the sample address space would be free. A single bit could be used to flag the first block in a new timecode frame, and two further bits would represent the frame rate. The last bit could be used to indicate that the timecode was asynchronous with the audio. The additional byte to indicate the offset in number of samples between the start of that block and the start of the frame (max. 191 samples) could be placed in the byte 5 position of channel status, since this is currently unused.

References

1   AES (1991) *AES11-1991 (ANSI 4.44-1991). Synchronization of digital audio equipment in studio operations*

2   AES (1984) AES5-1984. AES recommended practice for professional digital audio applications employing pulse code modulation – preferred sampling frequencies. *Journal of the Audio Engineering Society*, vol. 32, pp. 781–785

3   SHELTON, W.T. (1989) Synchronization of digital audio. In *Proceedings of the AES/EBU Interface Conference*, 12–13 September, London, pp. 92–116, Audio Engineering Society British Section

4   DUNN, N.J. (1992) Jitter: specification and assessment in digital audio equipment. Presented at the *93rd AES Convention, San Francisco*, 1–4 October, preprint no. 3361 (C-2)

5   DUNN, N.J. (1991) Considerations for interfacing digital audio equipment to the standards AES3, AES5 and AES11. In *Proceedings of the AES 10th International Conference*, 7–9 September, pp. 115–126

6   STORY, M. (1989) The AES interface and analog–digital conversion. In *Proceedings of the AES/EBU Interface Conference*, 12–13 September, London, pp. 16–21, Audio Engineering Society British Section

7   DUNN, C. and HAWKSFORD, M.O. (1992) Is the AES/EBU/SPDIF digital audio interface flawed?. Presented at the *93rd AES Convention, San Francisco*, 1–4 October, preprint no. 3360 (C-1)

8   GILCHRIST, N.H.C. (1980) Sampling-rate synchronization of digital sound signals by variable delay. *EBU Technical Review*, no. 183, October

9   PARKER, M. (1991) Sample frequency conversion, sample slippage, pitch changing and varispeed. In *Proceedings of the AES 10th International Conference*, 7–9 September, p. T-69

10 SHELTON, W.T. (1991) Timing inter-relations for audio with video. In *Proceedings of the AES 9th International Conference*, 1–2 February, pp. 31–44

11 BENSBERG, G. (1992) Time for digital audio within television. Presented at the *92nd AES Convention, Vienna*, 24–27 March, preprint no. 3258

12 KOMLY, A. and VIALLEVIELLE, A. (1990) *Synchronization and time codes in the user channel*. Proposal submitted to AES SC 2-5-1 working party on synchronization, Paris, March

13 EVANS, P. (1990) Digital audio in the broadcast centre. *EBU Technical Review*, no. 241/242, June–August

14 RUMSEY, F. J. (1992) *Timecode in channel status*. Proposal submitted to AES SC 2-5-1 working party on synchronization, San Francisco, August

Chapter 7

Practical audio interfacing

Despite the many standards relating to audio interfaces, or perhaps because of them, it is possible that practical difficulties may arise when attempting to interconnect two or more devices. This is the less than trivial problem of 'getting devices to talk to each other', and such problems may usually be boiled down to one of a few common sources of incompatibility. Not only must the user know how to work around basic incompatibilities, but also it is necessary to be aware of the dangers of incorrect communication – since devices may be made to 'talk' but something may be lost or gained in the translation!

The majority of this chapter is devoted to the standard two-channel interface. Communications between devices using dissimilar manufacturer-specific interfaces such as those described in Chapter 5 will nearly always require the use of a format convertor such as those described in section 7.6.1. In addition to the extended discussion of practical interfacing contained in the main text, there is a reference 'troubleshooting guide' to audio interfacing at the end of this chapter.

7.1 Incompatibilities between devices using the standard two-channel interface

In this section we shall look at the problems which may arise when using nominally the same interface at both ends of the link, discussing also the question of consumer-to-professional communications and vice versa.

There are only really two possible reasons why devices will not communicate, summed up as either electrical incompatibility or data incompatibility. If the two devices are using the same electrical interface, such as AES professional standard on XLR connectors or the IEC 958/EIAJ CP-340 consumer standard on phono connectors, then there is only a very small chance of electrical mismatch and the more likely cause of any problems is that differences exist in the data transmitted from that expected by the receiver. If direct links are attempted between consumer and professional equipment then there is potential for both electrical and data incompatibility, as discussed below.

7.1.1 Electrical mismatch in professional systems

Between identical interfaces (transmitter and receiver) the most likely electrical problems to arise are (a) loss over long lines; (b) noise and distortion over long lines;

(c) impedance mismatch. These can in general be avoided by good system design and by adhering to the recommendations contained in the appropriate standard, but occasionally one encounters a poor electrical installation or needs to make use of existing wiring which may not be ideal for the job of carrying digital audio. When such electrical problems are encountered the most likely symptom is that the receiver will find it difficult or impossible to lock to the incoming data signal, resulting in intermittent operation or indication of 'loss of lock' at the digital input. In practice receivers vary widely in their ability to lock to poor quality signals, and thus it may be found that a signal which works satisfactorily with one receiver proves unsatisfactory with another. Using devices such as those discussed in section 7.3 it is possible to determine the 'health' of the received data signal, as well as examining problems with data.

A receiver conforming to AES3 should be able to decode a signal with a minimum eye height of 200 mV, and since the transmitter normally produces at least 5 volts the resistive cable attenuation has to be quite large before it will reduce the eye height below this value. More problematical than simple resistive loss is high frequency roll-off over a long cable and this may need to be corrected for by using suitable equalization at the receiver (see section 4.2.3), although equalization should be treated with care since it will only work if the line is relatively noise free. (If the line is noisy then equalization may actually make the problem worse, and thus it has been suggested that if such equalization is to be used it perhaps belongs at the transmitter end rather than the receiver.)

High frequency loss affects the narrow pulses of the data stream before the wide ones, and these narrow pulses may either fail to provide a zero crossing or even disappear altogether in extreme cases. The greater the HF loss the more likelihood there is of intersymbol interference and data edge timing jitter, making it more difficult for the receiver to lock to the received data. Dunn[1] suggests that the cable used should not exhibit attenuation at 6 MHz which is more than 6 dB greater than that at 1 MHz over the distance used, otherwise equalization will be required. Although conventional analog audio cable can and has been used in many cases, it is becoming common for new installations to be wired for digital audio with cable having higher specifications and a characteristic impedance which is better controlled and closer to 110 ohms than conventional microphone cable.

Impedance mismatches were more likely under the original AES3 specification than they are under the 1992 revision, due to the 250 ohm termination impedance specified in AES3-1985. (This was to allow for between one and four receivers to be connected in parallel across one signal line.) Such mismatches may result in internal reflections such that transmitted pulses are reflected from the receiving end to interfere with pulses travelling in the opposite direction, and the cable may begin to function as an antenna – both picking up and radiating interference. Now that the termination impedance is specified at 110 ohms, and point-to-point interconnects are required, it may be necessary to modify the input impedance of older receivers by fitting a parallel resistor of around 200 ohms so as to make the termination nearer to the correct value. If it is necessary to feed more than one receiver from a single driver it is recommended that suitable digital distribution amplifiers (DDAs) are used, or, alternatively, one could use passive resistive splitters, although the DDA is preferable.

In order to avoid impedance mismatches in between transmitter and receiver it is important that the cable used is consistent along its length, and that joins are not made between dissimilar cable types. Problems may also arise if digital audio signals are routed via analog patch bays in which short sections of cabling with mismatched impedances may exist. Any cable 'stubs' in such installations produce short-delayed reflections with fairly high amplitude which can interfere with the transmitted data signal.

Users wishing to carry signals over long distances may wish to consider the possibility of adopting the 75 ohm unbalanced interconnect which is proposed as an alternative to balanced 110 ohm interfacing (although not yet standardized). Convertors are available which perform the job very easily. It has been suggested by a number of sources that HF interference such as RF sources is better rejected by the effectiveness of cable screening than by the balance of the electrical interface, and that 75 ohm coax cable has better controlled impedance than audio microphone cable.

A final point to bear in mind is that although transformers are not mandatory in most versions of the two-channel interface standard they may be used to ensure good electrical isolation and earth separation where appropriate. The transformer is standard in the EBU version of the interface, since it was regarded as important in broadcast studio centres where earth continuity is normally avoided between operational areas.

7.1.2 Data mismatch in professional systems

Data mismatch between professional devices using the standard two-channel interface is normally confined to problems with the implementation of channel status (see section 4.7), but there is also the possibility for differences in sampling rate and audio wordlength, as discussed in section 7.2, below. In future it is possible that user bit incompatibilities will also arise, as greater use is made of this channel, but few current professional devices take any notice of the state of the user bit. The validity bit is another potential root of trouble, and its handling is discussed in section 4.5. As discussed in section 7.6, a number of manufacturers now produce devices specifically designed to analyse and/or correct for data incompatibilities between pieces of equipment, and such 'fix-it' boxes can be very useful in encouraging communications between systems.

Incompatibilities in channel status implementation can give rise to a variety of symptoms ranging from complete failure of communication to seemingly correct but actually improper communication. The reason that such incompatibilities have arisen is largely due to the fact that the original AES3 specification was less than specific on how devices should set channel status bits that were not used, and how they should respond when receiving data that they were incapable of handling. Consequently all sorts of channel status implementations exist in commercial products, although it should be said that most of the time devices communicate without problems. Because of these potential difficulties, AES3-1992 is more specific about channel status implementation, specifying three levels of implementation depending upon the application (see section 4.7.4).

The most common problem areas in channel status implementation are (a) in the CRCC byte (byte 23); (b) in the signalling of pre-emphasis; (c) in the setting of the consumer/professional flag; and (d) in the indication of sampling rate. Less common problems may arise whereby the left and right channels have different channel status data (making it difficult to decide which is correct) or where the channel status block is the wrong length (one famous example exists of a 191 byte channel status block!).

The problem with the CRCC byte is that not all transmitting devices include it at the end of the channel status block, and some early devices implemented it incorrectly, thus confusing devices which expect to see the correct CRCC data. Receivers which check the CRCC will flag an almost continuous CRC error when decoding an input signal which does not contain CRCC or where it is incorrectly implemented, and the reaction of such a receiver will vary from complete refusal to accept the signal to acceptance of the signal whilst flagging a CRC error. It is impossible to state what the 'correct' response should be in such a case, since there is no way of telling whether the channel status data is correct or not, and thus it might be suggested that it should be left up to the user to select the appropriate response. Interface signal processors such as the 'fix-it' boxes mentioned above often perform the useful function of inserting correct CRCC data into the channel status blocks of signals which lack it. It might reasonably be assumed that a device seeing CRCC bytes repeatedly set to zero would assume that it was not in use and thereafter ignore it, but this is rarely the case at present.

The type of pre-emphasis used should be signalled in bits 2–4 of byte 0 of channel status, and it is possible that emphasis may have been applied to a signal without signalling this in channel status. Such incorrectness will not prevent the interface from working but may give rise to a pre-emphasized signal being carried through further stages in the signal chain without being de-emphasized. The only way to tell if a signal is pre-emphasized (when it is not indicated) is really to listen to it, since pre-emphasized signals will have an exaggerated HF response. The correct pre-emphasis flags may be set using a suitable interface processor, or the signal may be de-emphasized in the digital domain and the flags set to the no emphasis state.

Apart from the consumer/professional flag (discussed below), the sampling rate indication is the other main area of difficulty in channel status. Not all devices indicate the sampling rate of the signal in bits 6 and 7 of byte 0, and this can cause a lack of communication when received by a device expecting to see such data. Japanese devices in particular often will not accept data if it does not have the sampling rate flags set correctly. It is also difficult for a receiver to know what to do when presented with a sampling rate flag that contradicts the true audio sampling rate. In such cases it might be suggested that the device should rely on its detection of the true rate rather than the indicated rate, whilst perhaps flagging a mismatch on the front panel. Some confusion also exists over the interpretation of bit 5, byte 0 which indicates 'source sampling frequency unlocked'. The question is 'with reference to what is the sampling frequency unlocked?' – the internal clock, an external reference? When this bit is set to '0' (the default state) nothing can be concluded about the locked state of the sample clock, and many systems do not set this bit anyway. When set to '1' all one can say is that there is some problem with the lock of the sample clock and that its frequency may not be relied upon. In systems synchronized to a master reference signal its presence would be used to indicate that there was a free-running clock in a source device earlier in the signal chain.

7.1.3 Electrical mismatch between consumer and professional systems

Although it is not explicitly stated in IEC 958, the unbalanced electrical interface terminating in phono connectors is the one normally found on consumer equipment, and tends to be referred to as 'IEC 958 Type 2'. It is also commonly referred to as 'SPDIF', and very occasionally as 'CP-340 Type 2'. Thankfully, EIAJ CP-340 is more specific than IEC 958 in stating that this interface applies only to Type 2 (consumer) data.

As described in Chapter 4 there is so much similarity between the consumer and professional interfaces that it is tempting to think that consumer devices can be connected directly to professional systems or vice versa. Indeed there is a strong operational motive for this, since consumer and professional digital equipment are often mixed in studios and program material is often copied between systems. The problem is that although it is possible in many cases to make the electrical interconnection work, there are other difficulties to contend with such as the almost total dissimilarity in channel status and user bits, as well as the possibility that the consumer device may have a much less stable sample clock and be unable to lock to an external reference, giving rise either to problems in decoding or the danger that sample clock jitter will be passed on to other devices in the professional system. Ideally, therefore, one should not countenance the direct interconnection of consumer and professional equipment, preferring rather to use one of the many interface convertors that exist on the market which will set the necessary channel status bits correctly and convert the signal to the new electrical format. It may also be necessary to resynchronize the signal from a consumer device using a buffer store or sample rate convertor (see section 6.5) by locking it to the reference signal of the professional system, thus ensuring that its sampling rate is the same as that of the professional system and hopefully removing any excessive clock jitter.

Since operational situations are not usually ideal, and one may not have access to suitable format convertors and resynchronizers, it may be necessary to attempt direct interconnections and thus some guidelines will be given. Clearly such set-ups should be viewed as temporary, and accepting the possible incompatibilities in channel status (see below). Consumer-to-professional electrical connection is often possible because the consumer interface peak output voltage is around 0.5 V and the minimum allowed input voltage to a professional system is 0.2 V, and therefore provided that the wire is not too long the signal may be decoded. As shown in Figure 7.1 the centre core of the consumer coaxial lead may be connected to pin 2 of the professional XLR, and the shield to pins 1 and 3. Clearly there will be an impedance mismatch, and commercial impedance transformers are available to convert between either 75 and 250 ohms or 75 and 110 ohms. One Japanese manufacturer recommends the circuit shown in Figure 7.2 to balance a consumer output for carrying it over longer distances.

Professional-to-consumer connection may also work, again depending on channel status, since a consumer input is not normally damaged by the higher professional signal voltage. Two circuits are shown in Figure 7.3, depending on whether the professional output is transformer balanced (floating) or driven directly from TTL-level chips (balanced, but not floating). It is also possible to convert signals between consumer electrical and optical formats. Suggested circuits are shown in Figure 7.4.

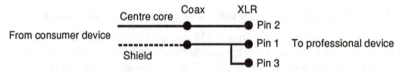

Figure 7.1 Temporary method of interconnection between consumer and professional interfaces

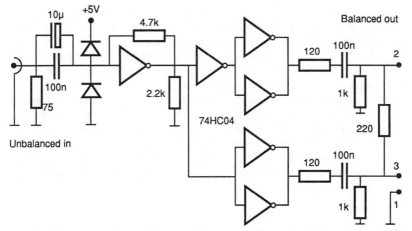

Figure 7.2 An example of a circuit suggested by one manufacturer for deriving a balanced digital output from consumer equipment

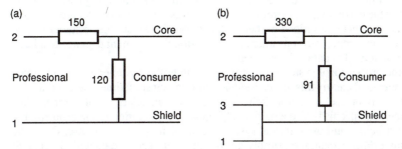

Figure 7.3 Examples of impedance matching adaption circuits between professional and consumer devices: (a) non-floating, and (b) floating (transformer-balanced) professional sources

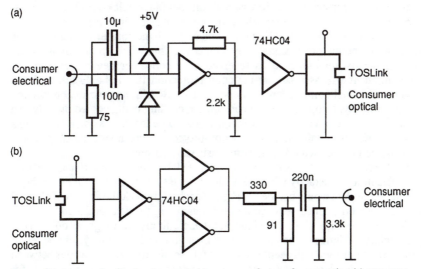

Figure 7.4 An example of a circuit suggested by one manufacturer for converting (a) consumer electrical to consumer optical format, and (b) optical to electrical

Unfortunately IEC 958 does not exclude the possibility of using the Type 2 *electrical* interface with professional *data*, or indeed vice versa (the two subjects are entirely separate in the document), and this has occasionally led to some unexpected implementations. Although one normally expects to find professional data coming out of XLR connectors there are cases where manufacturers or dealers have simply taken a consumer device and provided it with a so-called 'professional' output by feeding the consumer output via an RS-422 driver to an XLR connector. This may be cheap but it is to be discouraged since it leads users to think that the equipment may be connected directly to professional systems, whereas in practice the professional system may refuse to accept it due to the consumer flag being set in the first bit of channel status. If the data is accepted it is possible that further difficulty may arise due to the misinterpretation of channel status data (see section 7.1.4).

7.1.4 Data mismatch between consumer and professional systems

Although the audio part of the subframe is to all intents and purposes identical between the two interfaces there are key differences in the channel status data which have already been described in theory in Chapter 4. The success or otherwise of directly interconnecting consumer and professional equipment depends to a large extent on how this data is interpreted.

Some professional systems are provided with both consumer and professional interfaces, and clearly this offers the ideal solution, but sometimes there is only one electrical interface with a switch to select 'consumer' or 'professional' data characteristics. This is also a moderately satisfactory solution to the problem, but it still depends on how fully the system has implemented either case. Often the 'consumer' implementation simply means that it will accept data with the first bit of channel status set to zero, ignoring most or all of the following data.

Looking back to Figures 4.15 and 4.18 it will be seen that the similarity in channel status between consumer and professional runs out fairly quickly. One of the most common problems is in the signalling of pre-emphasis, since there are 3 bits to indicate this in the professional standard and only 2 in the consumer standard – the additional bit being used to signal copy protection. Thus when a professional signal with 'emphasis not indicated' is received on a consumer device, without any intervention, it will assume that the copy protection flag is being asserted and prevent the signal being recorded (depending on SCMS, as described in section 4.7.7). Similarly, a copy-protected consumer recording being copied directly to a professional system without intervention would force the professional device to a 'no emphasis' state, although a non-copy-protected consumer recording would normally be OK and would set the emphasis state correctly.

Past the emphasis bits there is no similarity at all between the interfaces, and thus it is difficult to say exactly what the results of one system interpreting the other's channel status data would be. Ideally a receiver should check the first bit of channel status to detect the consumer or professional nature of the data, and then switch its implementation automatically to interpret it accordingly, although, as pointed out quite rightly by Finger[2], the aims of consumer and professional system development are different and will become increasingly so, making the disparities between channel status implementation ever wider as this section of the subframe is used for more diverse applications.

7.2 Handling differences in audio signal rate and resolution

Because of the variety of sampling rates in use in digital audio and the increasing use of audio sample resolutions beyond 16 bits it is important to ensure that the audio signal retains optimum sound quality when it is converted digitally from one rate or resolution to another. The question of sample rate conversion has already been covered to some extent in Chapter 6, since it is closely related to the topic of synchronization, and there is little more to be said here except to state that normally it is impossible to interconnect two devices digitally whose sampling rates differ from one another, requiring that a sample rate convertor be used between the two. The question of differences in sample wordlength, though, will be covered in more detail.

The standard two-channel interface allows for up to 24 bits of audio per sample, linearly encoded in fixed point form. Until recently only 16 of these were normally used, with the remaining bits set to zero and the MSB of the 16 bit sample in the bit 27 position, but the question now arises as to how to cope with signals of, say, 18 or 20 bit resolution when they are digitally connected to devices of lower resolution. A number of techniques can be used to adjust the gain of, say, a 20 bit signal if necessary to reduce its resolution to 16 bits. These range from straightforward truncation, through bit-shifting and re-dithering at the new resolution, to developments which involve intelligent rounding of the truncation error by noise shaping. In future it may be that professional digital audio devices will incorporate internal intelligent procedures to handle signals of a higher resolution than their internal architecture allows, but at the moment it is normally necessary to employ external processing of some kind at such a juncture.

Truncation is the worst possible solution, and involves simply losing the least significant bits of the word. Without re-dithering the result of truncation is very unpleasant low-level distortion. If a 20 bit source were connected digitally to a 16 bit destination without any intermediate processing the result would normally be the straightforward truncation of the four LSBs.

The addition of dither noise in the digital domain at the point where resolution is reduced is a suitable means of improving the distortion situation, and this has been implemented on some digital interface processors and in professional digital mixers. Some editors also have various dithering algorithms for this purpose. The process randomizes the quantizing error that results from truncation by adding a pseudo-random number sequence of controlled amplitude and spectrum to the incoming audio data. In addition, if the full dynamic range of the digital signal has not been used by the programme material (if headroom has been left, for example) it may be possible to bit-shift the 20 bit samples upwards before truncating and re-dithering, such that more of the MSBs are used, thus losing less of the information contained in the LSBs. This is achieved by a simple increase of gain in the digital domain prior to 16 bit transfer.

In the 1992 revision of the AES3 interface standard, provision is made for much more careful definition of the sample wordlength, such that receiving devices may optimize the transfer of data from a transmitter of different resolution, as discussed in section 4.7.3. Standardization work has also gone on within the EBU to determine how analog signal levels should relate to digital signal levels, especially since 20 bit recording can be used on the audio tracks of digital video recorders. The conclusion reached was that one could not rely on correct implementation of byte 2 of channel status in devices using the AES/EBU interface in all cases, especially in older equipment, and thus that, irrespective of the number of bits, the only practical argument was for a fixed relationship to be used between analog and digital levels[3].

Originally EBU Recommendation R-64 specified that analog alignment level (corresponding to a meter reading of PPM4 or 0 dBu electrically) should be set to read 12 dB below full scale (i.e. −12 dB FS) on a digital system. This was based on the dynamic range available from typical 16 bit convertors, and assumed that the finished programme's level would be well controlled. Since 16 bit convertor technology has improved, and since it was necessary to use the same alignment for 20 bit systems, the new draft recommendation now specifies alignment level to be 18 dB below full scale. This allows for an additional 6 dB of operational headroom.

7.3 Analysing the digital audio interface

Because of the wide variety of implementations possible in the standard interfaces, and because of the need to test digital interface signals to determine their 'health' as electrical signals, there have arisen a number of items of test equipment which may be used to analyse the characteristics of the signal. The following is just a short summary of testing techniques, more detailed coverage of which may be found in Cabot[4], Blair et al.[5], Mornington West[6], and Stone[7].

7.3.1 Eye pattern and pulse-width testing

The eye pattern (see sections 3.2.4 and 4.2) of a two-channel interface signal is a rough guide to its electrical 'health', and gives a clue as to the likelihood of it being decoded correctly. In a test set-up described by Cabot[4], pictured in Figure 7.5, it is possible to vary the attenuation and high frequency roll-off over a digital link so as to 'close the eye' of a random digital audio signal to the limits of the AES3 specification. Having done this one may then verify whether a receiver correctly decodes the data, and thus whether it is within the specification. The testing of eye

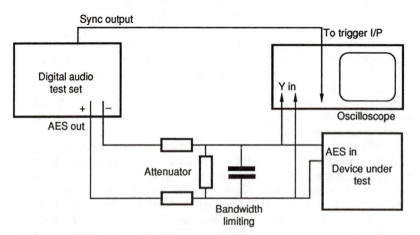

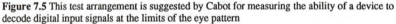

Figure 7.5 This test arrangement is suggested by Cabot for measuring the ability of a device to decode digital input signals at the limits of the eye pattern

patterns requires a high quality oscilloscope whose trigger input is derived from a stable clock locked to the source clock rate, preferably at a sub-multiple of it. Alternatively, provided that the clock recovery in the receiver is of high quality and rejects interconnect jitter it may be possible to use this, with the proviso that the eye pattern's reliability will be affected by any instability in the trigger signal.

An alternative to eye pattern testing on an oscilloscope is the use of a stand-alone interface analyser such as that described by Blair *et al.*[5] which is capable of displaying the amplitude and 'pulse smear' of the signal on an LCD display with relation to either an internal or external reference signal.

It has been suggested by Kondakor[8] of the BBC Designs and Equipment Department that eye patterns may not always indicate the 'decodability' of a signal, since he has found examples of seemingly good eye patterns which still give problems at the receiver. Consequently a device has been built which measures the variations in pulse width of received signals – a parameter which becomes greater on poor electrical links – displaying the result on a simple visual display which indicates variations in the three possible bit-cell timings (the '1', '0' and preamble wide pulses) in the form of bar lengths. It is claimed that this gives quick and easy verification of the reliability of received signals, and correlates well with the likelihood that a signal will be decoded.

7.3.2 Security margin estimation

Often it is necessary to get a rough idea of how close an interface is to failure, and one way of doing this is to introduce a broadband attenuator into the link at the receiving end (with sufficient bandwidth to accommodate the digital audio signal), gradually increasing the attenuation until the receiver fails to decode the signal. The amount of attenuation which can be introduced without the link failing is then a rough guide to the margin of security. An alternative to this method is for test equipment to examine the narrow pulses which follow the extra wide pulses in the subframe preambles of the data, since, due to intersymbol interference, these are usually the first to be reduced in width and level or even disappear in cases of poor links and limited link bandwidth. Some interface receiver ICs use this criterion as a measure of the security margin in hand.

7.3.3 Error checking

Digital interfaces are designed to be used in an error-free environment – in other words, errors are not anticipated over digital links, and there is no means of correcting for them. This is an achievable situation in well-designed systems, and above a certain signal-to-noise ratio one may expect no errors, but below this ratio the error rate will rise quite quickly. The error rate of a digital interface can be checked by a number of means. Common to all the techniques is that a certain bit pattern is generated by the test equipment and the received pattern is then compared against the generated pattern to check for data errors. It is then possible to attempt such things as increasing the noise level on the interface by injecting controlled levels and bandwidths of artificially generated noise to see the effect on error rates. A novel means of checking for errors also described by Cabot is to use a sine-wave test signal and to monitor the digital THD+N (distortion plus noise) reading at the receiver on suitable test equipment. Any error gives rise to a spike in the signal, resulting in a momentary increase in distortion which can be monitored at the output of the notch filter. He points out that errors in LSBs will produce less of an effect on the THD+N reading than errors in MSBs. By correlating errors with the data pattern at the time of the error it is possible to determine whether they are data dependent.

7.3.4 Other tests

Further checks on the interface may be performed on commercial test equipment, such as the accuracy of the sample clock (compared against either an external or calibrated internal reference), measurements of data cell jitter and common mode signal level. Test equipment can also show the states of all the information in channel status and other auxiliary bits if required, many systems converting these into useful human-readable indication. Examples exist of systems which will analyse the incoming channel status and other data and modify it to suit the established requirements of the receiver in order to set up proper communications in problem situations.

7.4 Interface transceiver chips

A number of dedicated ICs are now available for receiving and transmitting standard two-channel interface data, many operating in either consumer or professional modes. Such chips are large scale integrated circuits (LSIs) which take care of low jitter clock recovery, automatic CRCC checking and generation, automatic sample rate selection and channel status control, among other features. Examples of such devices are the BBC 'AESIC' chip[9], the Motorola DSP 56401[10], the Sony CXD 8277Q and 8278AQ[11] and the Crystal Semiconductors CS8411 and CS8412. In recent chips the clock separation and synchronization is carefully controlled such that a certain amount of buffering is incorporated within the receiver, in order that the clock derived from the incoming interface signal is separated from the clock used to read the data out of the IF chip to the rest of the system. Provided that master clock synchronization is used (see Chapter 6) the two rates will be the same, although perhaps slightly phase shifted. The interface between two-channel transceiver chips and the internal signal processing of the equipment is normally in a serial form which can be accepted directly by DSP devices such as the Motorola DSP 56000.

The Sony chip mentioned above, by using a novel design of phase-locked loop, is capable of very wide range lock to incoming interface signals with particularly fast lock-in to the received clock, as well as high stability, which is important in the case of receiving varispeeded signals. (Similar techniques are used in some of the other chips.) It also incorporates a means of detecting slipped clocks, such as may happen in the case of excessive jitter in the incoming signal, causing the phase between the incoming clock and the reference (used to control the readout from the chip's output buffer) to jump temporarily outside the limits of the AES11 synchronization tolerances, using a form of hysteresis controlled slip detector in the clock synchronization. When a clock is slipped a sample is either skipped or held, and a flag is raised to indicate the skip or hold, whilst placing an error signal in a separate latch to indicate which sample is in error. Internal signal processing can then read the error data and the appropriate flags and determine a suitable interpolation process to reduce the audible effects of the error. In this way the receiver could even accept asynchronous signals and possibly provide a glitch-free monitoring output, rather than muting, by using internal signal processing to interpolate. Clearly any slipping or holding of samples is non-ideal, but there may be operational situations in which a smooth audio output is desired despite the possibility of errors.

Multichannel interfacing using MADI (see section 4.8) is based on the FDDI (Fibre Distributed Digital Interface) standard, and requires the use of so-called 'TAXI' chips designed by AMD (Advanced Micro Devices). As described in Zölzer and Kalff[12], the Am7968 and 7969 chips can be used to handle transmission and reception of the high bit rate data from either optical fibre or copper cables, transferring this data normally in parallel form to and from the internal signal processing of the device in question in order to achieve the high transfer rate necessary.

7.5 Routers and switchers

There are essentially two ways by which standard two-channel interface signals may be routed and switched in systems such as broadcast studio centres where a number of sources and destinations exist. One technique is to use a TDM (Time-Division Multiplexed) switcher, and the other is to use a conventional logic-based crosspoint router. Which is appropriate depends largely on the importance of 'silent' switching, such as might be required for 'hot' or 'on-air' applications, since in a router used simply for assigning signals between sources and destinations (the equivalent to an analog jackfield) it may not matter that discontinuities arise when routing is altered.

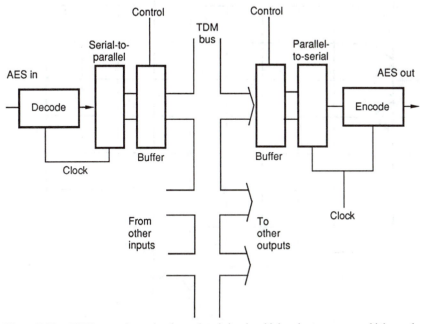

Figure 7.6 In a TDM router, input signals are decoded and multiplexed onto a common high speed bus. Inputs are routed to outputs by extracting data in the appropriate time slot for a particular input channel. Inputs must be synchronous for successful TDM routing

In a TDM switcher all the inputs must be synchronous and must be tightly phased inside the router to allow switching at precise timing points within the AES/EBU frame. The inputs are decoded and time-division multiplexed onto a fast parallel bus (as shown in Figure 7.6), allowing output routing to be determined by extracting data in the appropriate channel time slot on the bus and assigning it to a particular output. As suggested by de Jaham[13], the TDM switcher is appropriate when a large number of sources and destinations are involved because the physical space occupied 'per switching point' is smaller compared with a simple crosspoint router, yet the TDM switcher is complex in design and not usually cost effective for a small to medium number of inputs.

The crosspoint router may be a simple matrix of inputs and outputs, allowing any input to be routed to any output by selecting the appropriate 'crosspoint' between the two, using switching logic (see Figure 7.7). Furthermore, the signals may remain in the serial modulated form. One of the great advantages of the crosspoint router is that the inputs do not need to be synchronous, and it could even handle signals of different sampling rates. The problem with such a router is that the precise timing of the crosspoint may be difficult to control, and thus a glitch may result in the output signal at the time of switching. Evans[14] suggests that the typical level of such glitches is around −50 dBu in the analog domain, but this of course depends

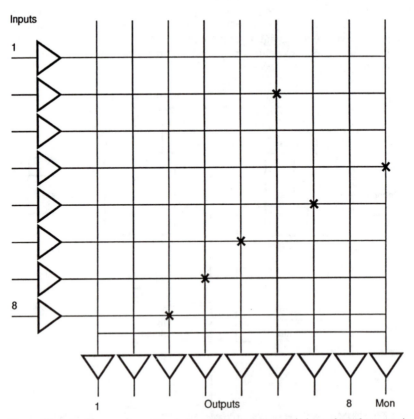

Figure 7.7 A simple crosspoint router may be used to connect a certain input channel to a certain output channel by selecting the appropriate crosspoint. Asynchronous signals may be handled here

on the relationship between analog and digital signal levels. Roe[15] indicates that clicks caused by such switching are usually masked by the digital filtering used in D/A convertors to a level up to around 50 dB below peak, and that a reframer may be used to detect and remove the corrupted sample from the data stream where noise-free switching is required. De Jaham also suggests that it is possible to mark the digital frames which have been corrupted at the switching point, such that they may be suitably concealed by equipment further down the signal chain, but points out that the process of marking is covered by patents belonging to TDF (Télédiffusion de France).

7.6 Other useful products

7.6.1 Interface format convertors

When operating in a mixed interface format environment, such as when wishing to copy a recording between a consumer and a professional system or when interfacing equipment using the Sony SDIF to equipment using AES/EBU, it may be necessary to employ a suitable format convertor. There are a number of such devices on the market, accepting inputs in one of a selection of standard formats (e.g. SDIF-2, SPDIF, AES/EBU, Mitsubishi, Yamaha) and converting them to one of a selection of output formats. Some such convertors also offer a number of intermediate signal processing operations, and a block diagram of one possible system is shown in Figure 7.8. As will be seen it offers high pass filtering to remove DC, the mixing of two stereo signal paths, the intelligent dithering of the signal for a chosen output resolution (16, 18 or 20 bits) and de-emphasis in the digital domain. Some more expensive systems also offer synchronization and sample rate conversion so that the system can transfer signals to systems of different sample rates, as well as different formats.

Often interface processors will take care of more subtle but extremely useful functions such as adding correct CRCC information to channel status, and checking for correct implementation of the standard. They may also allow the user to alter some or all of the channel status bits manually. With the advent of recording CD machines (CD-R) there have also arisen a number of systems capable of extracting the track start ID information from the user bits of the consumer digital interface, using this information to increment the track ID information of the destination format correctly, such as when a DAT tape is copied to a CD or vice versa. There is also the possibility that the processor may remove SCMS copy protection (see section 4.7.7) to allow professional users the flexibility that was lost with the introduction of the copy management system. Clearly such a device should be used with due regard to copyright laws.

7.6.2 Digital headphones

Once signals are interfaced digitally in a studio centre it becomes more difficult to monitor the signal audibly. In analog systems a pair of high impedance headphones can simply be plugged in to a jackfield to determine the presence or lack of a signal, but clearly this would not work in a digital system. There exist a small number of digital headphone products which contain an AES/EBU decoder and a built in D/A convertor, allowing the user to treat digital signals in a similar way to analog signals.

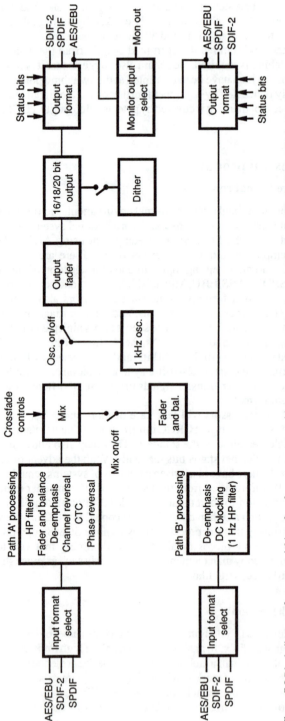

Figure 7.8 Block diagram of a commercial interface format convertor with some signal processing features. (Courtesy of Audio Digital Technology)

7.6.3 Digital distribution amplifiers (DDAs)

When a digital signal is to be routed to a number of destinations it is not advisable simply to parallel it. In order to avoid problems on decoding, as described at the start of this chapter, one should use a suitable DDA to split the signal. A DDA is also useful to distribute a source of the master reference signal to all devices in a studio.

7.7 A brief troubleshooting guide

If a digital interface between two devices appears not to be working it could be due to one or more of the following conditions. The reader should refer to the main text in the preceding chapters for more detailed explanation of the conditions described.

Asynchronous sample rates
The two devices must normally operate at the same sample rate, preferably locked to a common reference. Ensure that the receiver is in external sync mode and that a synchronizing signal (common to the transmitter) is present at the receiver's sync input. Alternatively, set the receiver to genlock to the clock contained in the digital audio input (standard two-channel interfaces only). If the incoming signal's transmitter cannot be locked to the reference it must be resynchronized or sample rate converted (see section 6.5).

'Sync' or 'locked' indicator flashing or out on the receiver normally means that no sync reference exists or that it is different from that of the signal at the digital input. Check that sync reference and input are at correct rate and locked to the same source. Decide on whether to use internal or external sync reference, depending on application.

If problems with 'good lock' or drifting offset arise when locking to other machines or when editing, check that any timecode is synchronous with video and sampling rate. If not, tape must be re-striped with timecode locked to the same reference as the recorder or a synchronizer used which will lock a digital audio input to the rate dictated by a timecode input.

Digital input
It may be that the receiver is not switched to accept a digital input.

Data format
Received data is in the wrong format. Both transmitter and receiver must operate to the same format. Conflicts may exist in such areas as channel status, and there may be a consumer–professional conflict. Use a format convertor to set the necessary flags.

Cables and connectors
Cables or connectors may be damaged or incorrectly wired. Cable may be too long, of the wrong impedance, or generally of poor quality. Digital signal may be of poor quality. Check eye height on scope against specification, and check for possible noise and interference sources. Alternatively make use of an interface analyser (see section 7.3).

SCMS (consumer interface only)
Copy protect or SCMS flag may be set by transmitter. For professional purposes, use a format convertor to set the necessary flags or use the professional interface which is not subject to SCMS.

Receiver mode
Receiver is not in record or input monitor mode. Some recorders must be at least in record–pause before they will give an audible and metered output derived from a digital input.

References

1   DUNN, J. (1991) Considerations for interfacing digital audio equipment to the standards AES3, AES5 and AES11. In *Proceedings of the AES 10th International Conference*, 7–9 September, p. 122, Audio Engineering Society

2   FINGER, R. (1992) AES3-1992: the revised two channel digital audio interface. *Journal of the Audio Engineering Society*, vol. 40, March, pp. 107–116

3   MØLLER, L. (1992) Signal levels across the EBU/AES digital audio interface. In *Proceedings of the 1st NAB Radio Montreux Symposium, Montreux, Switzerland*, 10–13 June, pp. 16–28, National Association of Broadcasters

4   CABOT, R. (1989) Measuring AES/EBU digital audio interfaces. *Journal of the Audio Engineering Society*, vol. 38, no. 6, October, pp. 459–467

5   BLAIR, I. *et al.* (1992) New techniques in analysing the digital audio interface. Presented at the *92nd AES Convention, Vienna*, 24–27 March, preprint no. 3230

6   MORNINGTON WEST, A. (1989) Signal analysis. In *Proceedings of the AES/EBU Interface Conference*, 12–13 September, pp. 83–91, Audio Engineering Society British Section

7   STONE, D. (1992) Digital signal analysis. *International Broadcast Engineer*, November

8   KONDAKOR, K. (1992) A pulse width analyser for the rapid testing of the AES/EBU serial digital audio signals. Presented at the *93rd AES Convention, San Francisco*, 1–4 October

9   BBC (1990) *AESIC: AES/EBU Interface Transceiver*, BBC Designs & Equipment Dept publication

10 MOTOROLA (1992) *DSP 56401: AES/EBU/CP340 digital audio transceiver*, Motorola Semiconductor Technical Data Sheet

11 SETOGAWA, T. (1992) Concept and technology of new AES/EBU DIO chips based on AES3-1992: for professional digital audio applications. Presented at the *92nd AES Convention, Vienna*, 24–27 March, preprint no. 3279

12 ZÖLZER, U. and KALFF, N. (1992) FDDI-based digital audio interfaces. Presented at the *92nd AES Convention, Vienna*, 24–27 March, preprint no. 3298

13 DE JAHAM, S. (1989) Asynchronous routing. In *Proceedings of the AES/EBU Interface Conference*, 12–13 September, pp. 64–68, Audio Engineering Society British Section

14 EVANS, P. (1990) Digital audio in the broadcast centre. *EBU Technical Review*, no. 241/242, June–August

15 ROE, G. (1991) Integrated routing in a hybrid environment. *International Broadcast Engineer*, July, pp. 44–46

Chapter 8

Digital video interfaces

In this chapter the various standardized interfaces for component and composite video will be detailed along with the necessary troubleshooting techniques.

8.1 Introduction

Of all the advantages of digital video, the most important for production work is the ability to record through multiple generations without quality loss. Effects machines perform transforms on images in the digital domain which remain impossible in the analog domain. For the highest quality post-production work, digital interconnection between such items as switchers, recorders and effects machines is highly desirable to avoid the degradation due to repeated conversion and filtering stages.

Video convertors universally use parallel connection, where all bits of the pixel value are applied simultaneously to separate pins. Rotary-head digital recorders lay data on the tape serially, but within the circuitry of the recorder, parallel presentation is more common because it allows slower, and hence cheaper, memory chips to be used for interleaving and timebase correction. The Reed–Solomon error correction depends upon symbols assembled from typically eight bits. Digital effects machines and switchers universally operate upon pixel values in parallel using fast multiplier chips.

Bearing all of this in mind, there is a strong argument for parallel connection, because the video naturally appears in the parallel format in typical machines. The first digital video interfaces were based on this reasoning. All that is necessary is a set of suitable driver chips, running at an appropriate sampling rate, to send video data down cables having separate conductors for each bit of the sample, along with a clock to tell the receiver when to sample the bit values. The cost in electronic components is very small, and for short distances this approach represents the optimum solution. An example would be where it is desired to dub a digital recording from one DVTR to another standing beside it.

Parallel connection has drawbacks too; these come into play when longer distances are contemplated. A multicore cable is expensive, and the connectors are physically large. It is difficult to provide good screening of a multicore cable without it becoming inflexible. More seriously, there are electronic problems with multicore cables. The propagation speeds of pulses down all of the cores in the cable will not be exactly the same, and so, at the end of a long cable, some bits may still be in transition when the clock arrives, while others may have begun to change to the value in the next pixel. In the presence of crosstalk between the conductors, and reflections due to sub-optimal termination, the data integrity eventually becomes marginal. Also the cables are physically larger than coax, making it difficult to install in existing plant.

Where it is proposed to interconnect a large number of units with a router, that device will be extremely complex because of the number of parallel signals to be handled. In short, parallel technology would never replace the central analog router of a television station. The answer to these problems is the serial connection. All of the digital samples are multiplexed into a serial bit stream, and this is encoded to form a self-clocking channel code which can be sent down a single channel. Skew caused by differences in propagation speed cannot then occur. The bit rate necessary is in excess of 200 Mbits/sec, but this is reasonably accommodated by coaxial cable. Provided suitable equalization and termination is used, it may be possible to send such a signal over cables intended for analog video. A similar approach was pioneered in digital audio by the AES/EBU serial interface, which allowed digital audio to be transmitted down existing twisted pair analog cabling (see Chapter 4).

The cabling savings implicit in such a system are obvious, but the electronic complexity of a serial interconnect is much greater, as high speed multiplexers or shift registers are necessary at the transmitting end, and a phase-locked loop, data separator and deserializer are needed at the receiver to regenerate the parallel signal needed within the equipment. Although this increased complexity raises cost, it has to be offset against the saving in cabling. The availability of specialized serial chips from a variety of manufacturers is speeding the acceptance of serial transmission.

A distinct advantage of serial transmission is that a matrix distribution unit or router is more easily realized. Where numerous pieces of video equipment need to be interconnected in various ways for different purposes, a crosspoint matrix is an obvious solution. With serial signals, only one switching element per signal is needed, whereas in a parallel system, a matrix would be unwieldy. A serial system has a potential disadvantage that the time distribution of bits within the block has to be closely defined, and, once standardized, it is extremely difficult to increase the wordlength if this is found to be necessary. The obsolete CCIR 8 bit serial system suffered from this problem, and has been replaced by a new serial digital interface standard (SDI) which is designed from the outset for 10 bit working but incorporates two low-order bits which can be transmitted as zero in 8 bit applications. In a parallel interconnect, the word extension can be achieved by adding extra conductors alongside the existing bits, which is much easier.

The third interconnect to be considered uses fibre optics. The advantages of this technology are numerous: the bandwidth of an optical fibre is staggering, as it is determined primarily by the response speed of the light source and sensor, and for this reason it has been adopted for a digital HDTV interface. The optical transmission is immune to electromagnetic interference from other sources, nor does it contribute any. This is advantageous for connections between cameras and control units, where a long cable run may be required in outside broadcast applications. The cable can be made completely from insulating materials, so that ground loops cannot occur, although many practical fibre-optic cables include electrical conductors for power and steel strands for mechanical strength.

Drawbacks of fibre optics are few. They do not like too many connectors in a given channel, as the losses at a connection are much greater than with an electrical plug and socket. It is preferable for the only breaks in the fibre to be at the transmitting and receiving points. For similar reasons, fibre optics are less suitable for distribution, where one source feeds many destinations. The bidirectional open-collector or tri-state buses of electronic systems cannot be implemented with fibre optics, nor is it easy to build a crosspoint matrix.

The high frequencies involved in digital video mean that accurate signal termination of electrical cables is mandatory. Cable losses cause the signal ampli-tude to fall with distance. As a result the familiar passive loop-through connection of analog video is just not possible. Whilst much digital equipment appears to have loop-through connections, close examination will reveal an amplifier symbol joining the input and output. Digital equipment must use active loop-through and if a unit loses power, the loop-through output will fail.

8.2 Characteristics of digital video signals

8.2.1 Video spectra

In video, sampling takes place in more than one dimension at once, and the resulting spectra can be complicated. It is well worth looking at the sampling spectra of analog video before the additional horizontal sampling of digital video takes place. In fact sampling theory can explain how the composite video standards are obtained.

For RGB working, a camera will produce three signals having the same scan rates and hence the same spectrum. In colour difference working one wideband lumi-nance signal having the same bandwidth as the components is accompanied by two colour difference signals which have the same spectral structure, but whose overall bandwidth is less. These three signals are sent on separate cables in analog

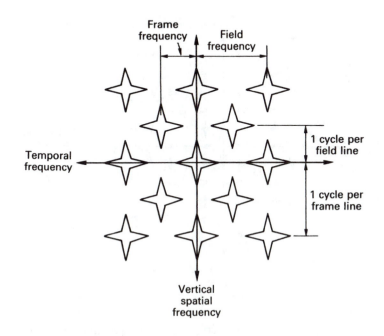

Figure 8.1 The vertical/temporal spectrum of monochrome video due to interlace

component interfaces. Three separate conversions to digital are required, but it will be seen that it is then possible to multiplex the three signals into one channel.

Figure 8.1 shows the spectrum of analog monochrome video (or of an analog component). The use of interlace has a pronounced effect on the vertical/temporal spectrum because of the quincunx sampling positions (resembling the five on dice). It will be seen that there are spaces within the spectrum which are apparently unused. These spaces will be occupied by the chroma signal, whose spectrum needs to be carefully tailored to reduce overlap with the luminance.

8.2.2 3-D spectrum of NTSC

Analog composite video can only be fully described in three dimensions because of the presence of the subcarrier. Digital video samples in three dimensions and also has a three-dimensional spectrum. As such spectra are difficult to illustrate, they are often portrayed in two dimensions at a time.

Figure 8.2 shows that the structure of the vertical/temporal spectrum of luminance is the same as that of the two colour difference signals because both have the same field and line rates. In NTSC (National Television Systems Committee), the three signals are encoded into a single channel having the same bandwidth as luminance by exploiting the periodic gaps in the spectrum. The two colour difference signals are used to generate reduced bandwidth signals which alternately phase modulate a subcarrier. The subcarrier is suppressed to produce a chroma signal[1].

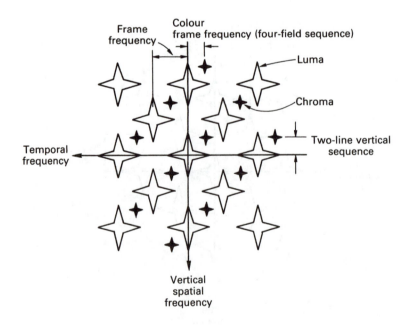

Figure 8.2 The vertical/temporal spectrum of NTSC. Note the one-quarter field rate component responsible for the four-field sequence

In order to allow the maximum perceived resolution with minimum bandwidth, the chroma phase was shifted slightly with respect to horizontal, so that subcarrier phase demodulation would give a pair of signals known as I and Q (in-phase and quadrature). This placed the Q signal on the colour diagram axis to which the eye is least sensitive. The chroma is given by:

$I \sin F_{sc} + Q \cos F_{sc}$

such that at 90 degree intervals, the chroma waveform represents I, Q, $-I$, $-Q$ repeatedly. If the receiver samples the chroma at the same intervals, but with the same phase shift, the two baseband colour difference signals can be restored.

The chroma modulation process takes the spectrum of the colour difference signals and produces upper and lower sidebands around the frequency of subcarrier. Since both colour and luminance signals have gaps in their spectra at multiples of line rate, it follows that the two spectra can be made to interleave and share the same spectrum if an appropriate subcarrier frequency is selected. The subcarrier frequency of NTSC is an odd multiple of half line rate: 227.5 times to be precise. Figure 8.3 shows that this frequency means that on successive lines the subcarrier will be phase inverted. There is thus a two-line sequence of subcarrier, responsible for a vertical frequency of half line frequency. The existence of line pairs means that two frames or four fields must elapse before the same relationship between line pairs and frame sync repeats. This is responsible for a temporal frequency component of half the frame rate. These two frequency components can be seen in the vertical/temporal spectrum of Figure 8.2.

The effect of the chroma added to luminance is to make the luminance alternately too dark or too bright. The phase inversion causes this effect to cancel over pairs of lines, giving compatibility with monochrome receivers. The half frame rate component is responsible for the familiar four-field colour framing sequence. When editing NTSC recordings, this four-field sequence must not be broken. If this rule is not followed, it is possible for a sudden inversion of subcarrier phase to occur at the edit point, which renders the signal unbroadcastable unless it undergoes additional processing.

As the chroma of NTSC is suppressed carrier, it is necessary for the receiver to regenerate the carrier frequency so that it knows when to sample the chroma. This is done by broadcasting a portion of subcarrier, known as the burst, between the sync pulse and the active line. A phase-locked loop in the receiver will freewheel between

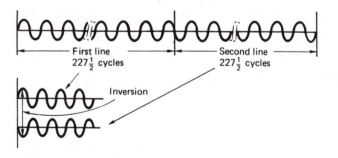

Figure 8.3 The half angle offset in F_{sc} results in a two-line sequence (a) where the first line ends on a half cycle, and the second line chroma will be inverted with respect to the first as at (b)

bursts. The half line offset of NTSC subcarrier means that if an oscilloscope is triggered from H-sync, normal and inverted bursts will appear superimposed on the trace.

Whilst the spectral interleaving of NTSC works fine for the video, there was a problem with the sound carrier. To prevent vision interference with the sound, the sound carrier of the 525/60 monochrome system had itself been placed at an odd multiple of half line rate, and so the addition of the chroma signal meant that chroma and sound could mutually interfere. The audio subcarrier frequency was standardized, and any change would have meant modification to every monochrome television set in the United States on the introduction of colour. The solution adopted was to contract the spectrum of the video slightly by reducing the frame rate to 29.97 Hz. This made the line rate 15 734.25 Hz and the subcarrier frequency 3.5795 MHz. The small disparity between field rate and 60 Hz was of little consequence when it was introduced, but the development of the video recorder revealed a difficulty in synchronizing NTSC recordings to real time, which was alleviated by the invention of drop-frame time code. This allows a single count to count real-time seconds and NTSC rate frames at the same time by having dropped frames, which do not exist in the video, but which make the seconds count correct. The use of February 29 in the calendar has a similar effect.

The dependence of NTSC on the absolute phase of subcarrier to accurately convey colour led to some difficulties in multipath reception conditions in high rise districts, and to the somewhat unfair redefinition of the system title as Never Twice the Same Colour. In fact NTSC works well in the majority of other locations.

8.2.3 3-D spectrum of PAL

When the PAL (Phase Alternating Line) system was being developed, it was decided that immunity to received phase errors should be a goal. Figure 8.4(a) shows how this was achieved. The two colour difference signals (U and V) are used to quadrature modulate a subcarrier in a similar way as for NTSC, except that the phase of the V signal is reversed on alternate lines. The receiver must then re-invert the V signal in sympathy. If a phase error occurs in transmission, it will cause the phase of V to alternately lead and lag, as shown in Figure 8.4(b). If the colour difference signals are averaged over two lines, the phase error is eliminated and replaced with a small saturation error which is subjectively much less visible. The receiver needs to know whether or not to invert V on a particular line, and this information is conveyed by swinging the phase of the burst ±135 degrees on alternate lines. A damped PLL in the receiver will run at the average phase, i.e. subcarrier phase, but it will develop a half line rate phase error whose polarity determines the sense of V-switch.

Figure 8.5(a) shows the vertical temporal spectrum of the U signal, which is identical to that of luminance. However, the inversion of V on alternate lines causes a two-line sequence which is responsible for a vertical frequency component of half line rate. As the two-line sequence does not divide into 625 lines, two frames elapse before the same relationship between V-switch and the line number repeats. This is responsible for a half frame rate temporal frequency component. Figure 8.5(b) shows the resultant vertical/temporal spectrum after V-switch and illustrates that the V component has shifted diagonally so that its spectral entries lie half-way between the U component entries. Spectral interleaving with a half cycle offset of subcarrier frequency as in NTSC will not work, as Figure 8.5(c) shows. The solution is to adopt

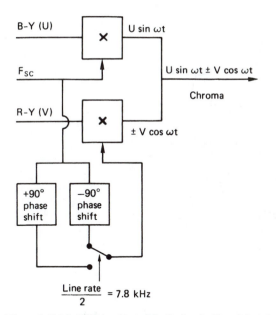

Figure 8.4(a) In PAL the phase of the feed to the V modulator is reversed on alternate lines

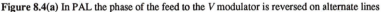
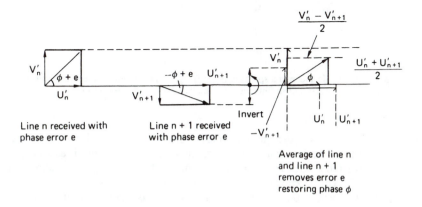

Chroma phase ϕ on line n

Chroma phase $-\phi$ on line n + 1

Line n received with phase error e

Line n + 1 received with phase error e

Invert

Average of line n and line n + 1 removes error e restoring phase ϕ

Figure 8.4(b) Top, chroma as transmitted. Bottom, chroma received with phase error. V is reversed on second line and the two lines are averaged, eliminating phase error and giving a small saturation error

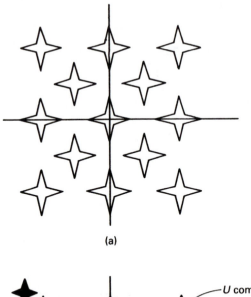

(a)

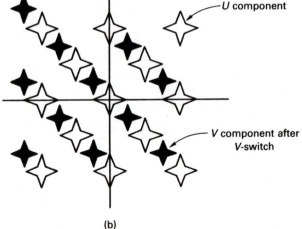

U component

V component after
V-switch

(b)

Figure 8.5 (a) The vertical/temporal spectrum of the U signal has the same structure as luminance. At (b) V-switch has the effect of shifting the spectrum diagonally with respect to U.

a subcarrier frequency with a quarter cycle per line offset[2]. Multiplying the line rate by 283.75 allows the luminance and chrominance spectra to mesh as in Figure 8.5(d). Note that there is an area of the spectrum which appears not to contain signal energy in PAL. This is known as the Fukinuki hole.

The quarter cycle offset is thus a fundamental consequence of elimination of phase errors in PAL, but it does cause some other effects which raise the complexity of implementation, leading to the retaliatory name of Problems Are Lurking.

The quarter cycle offset means that there are now line quartets instead of line pairs, and this results in a vertical frequency component of one-quarter of line rate

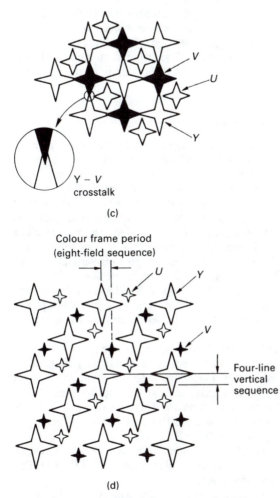

(c)

(d)

Figure 8.5 (contd.) A half cycle subcarrier offset will not work as (c) shows. Instead a quarter cycle offset is used as at (*d*)

which can be seen in the figure. Furthermore, four frames or eight fields have to elapse before the same relationship of subcarrier to frame timing repeats. This results in a temporal frequency component of one-quarter of frame rate which is also visible in the figure. This component restricts the way in which PAL recordings can be edited.

The three-quarter cycle offset of subcarrier also means that the line pair cancellation of NTSC is absent, and another means has to be found to achieve visibility reduction. This is done by adding a frequency of half frame rate to subcarrier frequency, such that an inversion in subcarrier is caused from one field to the next. Since in an interlaced system lines one field apart are adjacent on the screen, cancellation is achieved. The penalty of this approach is that subcarrier phase creeps forward with respect to H-sync at one cycle per frame. The eight-field sequence contains 2500 unique lines all having the subcarrier in a slightly different position. Observing burst on an H-triggered oscilloscope shows a stable envelope with a blurred interior. Measuring the phase of subcarrier with respect to sync requires specialist equipment or a great deal of determination. This is of little consequence for broadcasting, but it does raise the complexity of recorders and timebase correctors.

8.2.4 Choice of sampling rate: component

In principle, the sampling rate of a system need only satisfy the requirements of sampling theory and filter design. Any rate which does so can be used to convey a video signal from one place to another. In practice, however, there are a number of practical restraints which limit the choice of sampling rate considerably.

It should be borne in mind that a video signal represents a series of two-dimensional images. If a video signal is sampled at an arbitrary frequency, samples in successive lines will be in different places. If it is sampled at a rate which is a multiple of line rate the result will be that samples on successive lines will be in the same place and the picture will be converted to a neat array having vertical columns of samples. This allows for the vertical spatial filtering necessary in DVEs and standards convertors and for concealment in DVTRs. A line-locked sampling rate can be conveniently obtained by multiplying the frequency of H-sync using a phase-locked loop. The position of samples along the line is then determined by the leading edge of sync.

Whilst the bandwidth required by 525/59.94 is less than that required by 625/50, and a lower sampling rate might have been used, again practicality suggested a common sampling rate. The benefit of a standard H-locked sampling rate for component video is that the design of standards convertors is simplified and DVTRs have a constant data rate independent of standard. This was the goal of CCIR Recommendation 601[3], which combined the 625/50 input of EBU Doc. Tech. 3246 and 3247 with the 525/59.94 input of SMPTE RP 125.

CCIR 601 recommends the use of certain sampling rates which are based on integer multiples of the carefully chosen fundamental frequency of 3.375 MHz. This frequency is normalized to 1 in the document.

In order to sample 625/50 luminance signals without quality loss, the lowest multiple possible is 4 which represents a sampling rate of 13.5 MHz. This frequency line-locks to give 858 samples per line in 525/59.94 and 864 samples per line in 625/50. The spectra of such sampled luminance are shown in Figure 8.6.

In the component analog domain, the colour difference signals typically have one-half the bandwidth of the luminance signal. Thus a sampling rate multiple of 2 is used and results in 6.75 MHz. This sampling rate allows respectively 429 and 432 samples per line.

Component video sampled in this way has a 4:2:2 format. Whilst other combinations are possible, 4:2:2 is the format for which the majority of digital component equipment is constructed and is the only component format for which parallel and serial interface standards exist. The D-1, D-5 and Digital Betacam DVTRs operate with 4:2:2 format data. Figure 8.7 shows the spatial arrangement given by 4:2:2 sampling. Luminance samples appear at half the spacing of colour difference samples, and every other luminance sample is co-sited with a pair of colour difference samples. Co-siting is important because it allows all attributes of one

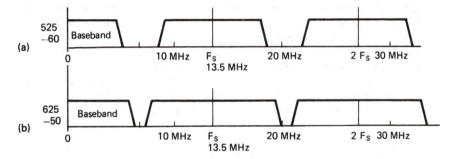

Figure 8.6 Spectra of video sampled at 13.5 MHz. At (a) the baseband 525/60 signal at left becomes the sidebands of the sampling rate and its harmonics. At (b) the same process for a 625/50 signal results in a smaller gap between the baseband and sideband because of the wider bandwidth of the 625 system. The same sampling rate for both standards results in a great deal of commonality between 50 Hz and 60 Hz equipment

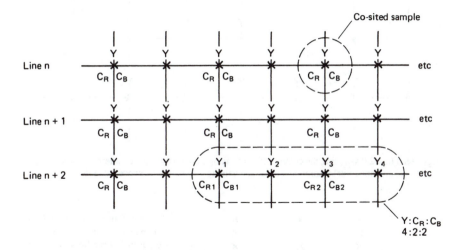

Figure 8.7 In CCIR-601 sampling mode 4:2:2, the line synchronous sampling rate of 13.5 MHz results in samples having the same position in successive lines, so that vertical columns are generated. The sampling rates of the colour difference signals C_r, C_b are one-half of that of luminance, i.e. 6.75 MHz, so that there are alternate Y-only samples and co-sited samples which describe Y, C_r and C_b. In a run of four samples, there will be four Y samples, two C_r samples and two C_b samples, hence 4:2:2

picture point to be conveyed with a three-sample vector quantity. Modification of the three samples allows such techniques as colour correction to be performed. This would be difficult without co-sited information. Co-siting is achieved by clocking the three ADCs simultaneously. In some equipment one ADC is multiplexed between the two colour difference signals. In order to obtain co-sited data it will then be necessary to have an analog delay in one of the signals.

For full bandwidth RGB working, 4:4:4 can be used with a possible 4:4:4:4 use if including a key signal. For lower bandwidths, multiples of 1 and 3 can also be used for colour difference and luminance respectively. 4:1:1 delivers colour bandwidth in excess of that required by the composite formats. 4:1:1 is used in the 525 line version of the DVC quarter-inch digital video format. 3:1:1 meets 525 line bandwidth requirements. The factors of 3 and 1 do not, however, offer a columnar structure and are inappropriate for quality post-production.

In 4:2:2 the colour difference signals are sampled horizontally at half the luminance sampling rate, yet the vertical colour difference sampling rates are the same as for luminance. Where bandwidth is important, it is possible to halve the

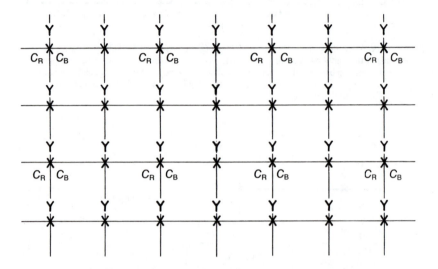

Figure 8.8 In 4:2:0 sampling the colour difference samples are only taken on alternate lines, making the vertical and horizontal resolution similar and halving the colour difference data rate

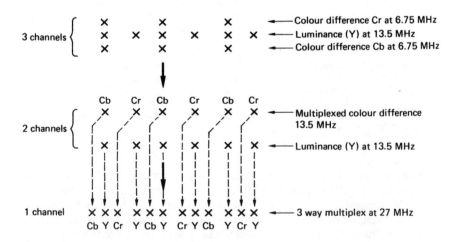

Figure 8.9 The colour difference sampling rate is one-half that of luminance, but there are *two* colour difference signals, C_r and C_b, hence the colour difference data rate is equal to the luminance data rate, and a 27 MHz interleaved format is possible in a single channel

vertical sampling rate of the colour difference signals as well. Figure 8.8 shows that in 4:2:0 sampling, the colour difference samples only exist on alternate lines so that the same vertical and horizontal resolution is obtained. 4:2:0 is used in the 625 line version of the DVC format and in the MPEG 'Main Level Main Profile' format for multimedia communications and, in particular, DVB.

Figure 8.9 shows that in 4:2:2 there is one luminance signal sampled at 13.5 MHz and two colour difference signals sampled at 6.75 MHz. Three separate signals with different clock rates are inconvenient and so multiplexing can be used. If the colour difference signals are multiplexed into one channel, then two 13.5 MHz channels will be required. Such an approach is commonly found within digital component processing equipment where the colour difference processing can take place in a single multiplexed channel.

If the colour difference and luminance channels are multiplexed into one, a 27 MHz clock will be required. The word order is standardized to be:

C_b, Y, C_r, Y, etc.

In order unambiguously to demultiplex the samples, the first sample in the line is defined as C_b and a unique sync pattern is required to identify the beginning of the active line.

There are two ways of handling 16:9 aspect ratio video. In the anamorphic approach both the camera and display scan wider but there is no change to the sampling rates employed and therefore the same 27 MHz data stream can be employed unchanged. Compared with 4:3, the horizontal spacing of the pixels in 16:9 must be greater as the same number are stretched across a wider picture. This results in a reduction of horizontal resolution, but standard 4:3 production equipment can be used subject to some modifications to the shape of pattern wipes in vision mixers. When viewed on a 4:3 monitor anamorphic signals appear squeezed horizontally.

In the second approach, the pixel spacing is kept the same as in 4:3 and the number of samples per active line must then be increased by 16:12. This requires the data rate to rise to 36 MHz. Thus the luminance sampling rate becomes 18 MHz and the colour difference sampling rate becomes 9 MHz. The D-5 DVTR can record such signals. Strictly speaking the format no longer adheres to CCIR-601 because the sampling rates are no longer integer multiples of 3.375 MHz. If, however, 18 MHz is considered to be covered by Rec. 601, then it must be described as 5.333...:2.666...:2.666... .

8.2.5 Choice of sampling rate: composite

When composite video is to be digitized, the input will be a single waveform having spectrally interleaved luminance and chroma. Any sampling rate which allows sufficient bandwidth would convey composite video from one point to another. However, if processing in the digital domain is contemplated, there will be less choice.

In the composite digital colour processor it will be necessary to decode the composite signal, which will require some kind of digital filter. Whilst it is possible to construct filters with any desired response, it is a fact that a digital filter whose response is simply related to the sampling rate will be much less complex to implement. This is the reasoning which has led to the near universal use of four times subcarrier sampling rate. Figure 8.10 shows the spectra of PAL and NTSC sampled

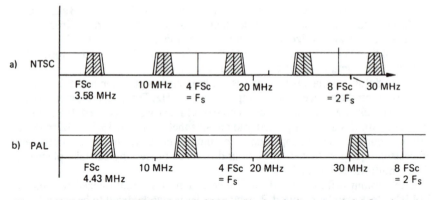

Figure 8.10 The spectra of NTSC at (a) and of PAL at (b) where both are sampled at four times the frequency of their respective subcarriers. This high sampling rate is unnecessary to satisfy sampling theory, and so both are oversampled systems. The advantages are in the large spectral gap between baseband and sideband which allows a more gentle filter slope to be employed, and in the relative ease of colour processing at a sampling rate related to subcarrier

at $4 \times F_{sc}$. It will be evident that there is a considerable space between the edge of the baseband and the lower sideband. This allows the anti-aliasing and reconstruction filters to have a more gradual cut-off, so that ripple in the passband can be reduced. This is particularly important for composite digital recorders, since they are digital devices in an analog environment, and signals may have been converted to and from the digital domain many times in the course of production. A subcarrier multiple sampling clock is easily obtained by gating burst to a phase-locked loop. In NTSC there is no burst swing, whereas at $4F_{sc}$, the burst swing of PAL moves burst crossings by exactly one sample period and so the phase relationship between burst crossings and $4F_{sc}$ clock is unaffected by burst swing.

In NTSC, siting of samples along the line is affected by ScH phase. In PAL, the presence of the 25 Hz component of subcarrier means that samples are not in exactly the same place from one line to the next. The columns lean over slightly such that at the bottom of a field there is a displacement of two samples with respect to the top.

8.2.6 Digital colour framing in PAL

In PAL, the eight-field sequence comes about because of the quarter cycle offset of subcarrier against line rate. There are 283.75 cycles of subcarrier in one line, which means that four lines must pass before the same relationship occurs between subcarrier and horizontal sync. The odd number of lines in a frame necessary for interlace mean that four frames must elapse before the four-line sequence assumes the same relationship with vertical sync. Unlike NTSC which has a half cycle offset, PAL does not naturally have good cancellation of chroma on the screen. As was seen in section 8.2.3, this is achieved by the addition of 25 Hz to the subcarrier frequency, which causes chroma on a given line to be anti-phase with chroma on the line in the next field which will be physically adjacent on the interlaced display. Although this 25 Hz component achieves its goal, it also complicates the colour framer because instead of having four different relationships of subcarrier to horizontal there are now 2500!

Since composite digital samples using burst as a phase reference, it is necessary to follow what the burst does with respect to horizontal. Owing to the quarter cycle offset, subcarrier phase advances by 90 degrees from line to line, but the burst swings forward 90 degrees on one line and back 90 degrees on the next to convey the state of *V*-switch. This means that the burst swing will alternately cancel or double the subcarrier phase advance, giving two lines with the same burst to sync relationship followed by two lines with an inverted burst to sync relationship. Figure 8.11 shows that these four lines repeat at 3.9 kHz and contain two complete cycles of *V*-switch which runs at 7.8 kHz. Once every four frames, line one of the frame will be the first of the four line types, and this fact will be indicated in a VTR by the colour framing flag on the control track. The composite SDI carries colour framing information in the form of a field count. It is the job of the colour framer to produce the colour framing flag and field count from input video.

The sampling clock in PAL uses burst as a phase reference. The colour framer must use sample values obtained by digitizing the leading edge of sync with a burst-phased sampling clock.

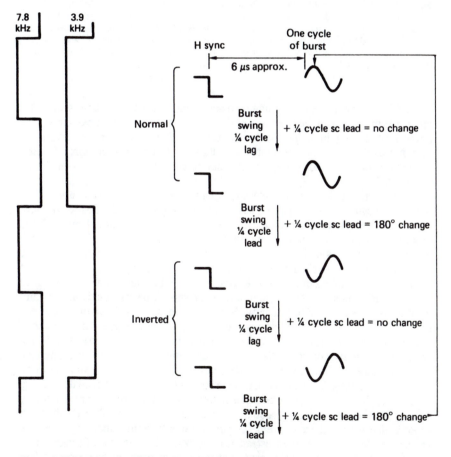

Figure 8.11 Quarter cycle subcarrier/H relationship and burst swing combine to give four-line sequence

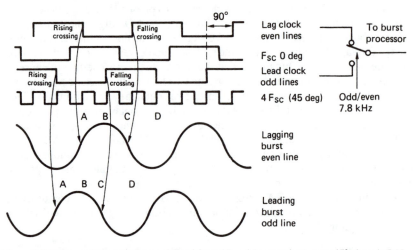

Figure 8.12 Two quadrature clocks are derived from $4F_{sc}$ whose crossings are at 45° phase in PAL. These two clocks are used on alternate lines, where they predict the position and direction of a burst crossing. If it is a downward crossing, the phase error needs to be inverted. If the odd/even sequence gets out of step, the clock moves forward a quadrant when an incoming burst moves back a quadrant, so the twist is perceived as inverted. The phase error alternates at half line rate and can be used to reset the 7.8 kHz divider

The first stage in colour framing is to determine the phase of burst swing. This process is integral with the operation of the sampling clock phase-locked loop. The phase-locked loop includes the video ADC so that chroma phase errors are minimized. The ADC runs constantly, and digitizes the entire video signal. Control of the VCO is achieved by examining the numerical value of samples taken during the burst.

Figure 8.12 shows one possible implementation. Owing to PAL burst swing, burst phase can have one of two values, 90 degrees apart, so both of them will have the same phase relationship to $4 \times F_{sc}$. A two-phase divider running from the VCO produces a pair of quadrature square waves at subcarrier frequency. These are used on alternate lines to synchronize the numerical processing of sampled burst. In the first quadrant, A, there should be a positive burst crossing, and blanking level, or 64 is subtracted from the sample value to give a phase error measurement. In the next quadrant, B, a positive burst value above 64 should be obtained. In the third quadrant, C, there should be a negative burst crossing, so 64 will be subtracted from the sample value, and the result inverted to give a phase error value. The phase error values for several cycles of burst are numerically averaged and fed to a DAC which produces the VCO control voltage. Upon first applying an input, the alternate selection of the quadrature control signals will have a random phase which may be right or wrong. If by chance it is correct, the above sequence of events will hold true for all lines. However, if the selection is in the wrong phase, the timing will lag by a quadrant when burst leads by a quadrant and vice versa. This will make the burst invert with respect to the measurement timing on alternate lines, and a negative peak value below 64 will be obtained instead of a positive peak value above 64 in quadrant B. This condition is easily detected, and is used to reset the phase of the line frequency divider. This divider will now be truly reflecting the input burst swing,

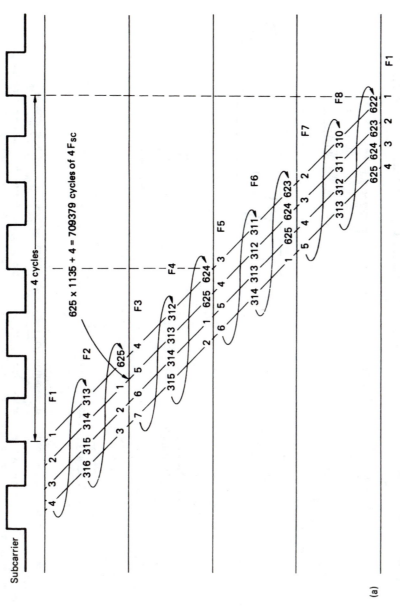

Subcarrier

4 cycles

625 x 1135 + 4 = 709379 cycles of 4 F_sc

F1

316 315 314 313

4 3 2 1

F2

625

315 314 313 312

7 6 5 4

F3

1 625 624

315 314 313 312

2 1

6 5 4 3

F4

1 625 624

314 313 312 311

3 2 1

5 4 3 2

F5

1 625 624 623

313 312 311 310

4 3 2

5 4 3 2

F6

F7

625 624 623 622

313 312 311 310

4 3 2 1

F8

F1

(a)

Figure 8.13(a) The position of sync with respect to subcarrier can be found for any line in an eight-field sequence using this chart. The quarter cycle offset causes sync to move forward one quadrant per line over a four-line sequence, but the addition of 25 Hz causes sync to slip back towards subcarrier by one cycle per frame

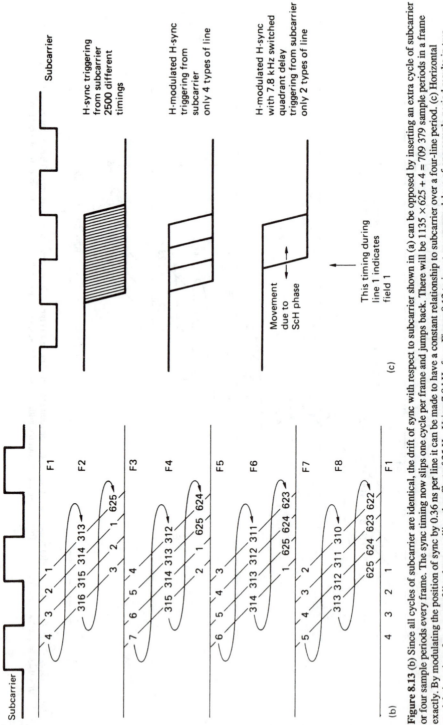

Figure 8.13 (b) Since all cycles of subcarrier are identical, the drift of sync with respect to subcarrier can be opposed by inserting an extra cycle of subcarrier or four sample periods every frame. The sync timing now slips one cycle per frame and jumps back. There will be $1135 \times 625 + 4 = 709\ 379$ sample periods in a frame exactly. By modulating the position of sync by 0.36 ns per line it can be made to have a constant relationship to subcarrier over a four-line period. (c) Horizontal modulation gives four types of line by cancelling the effect of 25 Hz. Using 7.8 kHz from Figure 8.12 to insert or remove delays of one sample period results in two relationships. Sampling during line 1 will determine whether field 1 or field 5 is present

and an accurate phase comparison can be made between burst and sampling clock on every line. The line frequency divider is now producing 7.8 kHz.

By numerically analysing the ADC output, it is easy to detect the sync pulses, and using programmable counters it is possible to detect both vertical and horizontal syncs from their duration. In this way a gate can be opened during the first line of each frame. During the gate, the state of 7.8 kHz is examined, and this will only be true on alternate frames, and allows four-field colour framing to be achieved. The burst swing (7.8 kHz) derived colour framing cannot tell whether it has found field 1 or field 5, and this can only be resolved by looking at the 3.9 kHz relationship between burst and sync.

The problems involved in determining the colour frame are summarized below, prior to the somewhat involved solutions.

Since sampling is subcarrier locked, it is necessary to consider how sync pulses move with respect to subcarrier rather than the other way round. The quarter cycle offset means that the position of a sync edge will move forward by one quadrant of subcarrier per line.

The presence of the 25 Hz offset causes sync edges to constantly slide back with respect to the sampling clock by four cycles or 225 ns per frame. The absolute position of the sync edge depends on ScH phase of the input signal which can drift.

Assessment of ScH phase according to the definition can only be made once every eight fields, and reliance on a single measurement is prone to noise. Techniques must be found to allow a colour framer to make a continuous assessment of framing so that a faster, more reliable lock is achieved.

Figure 8.13(a) shows how the phase of sync with respect to subcarrier can be found for any line in the eight-field sequence. The presence of the 25 Hz component is responsible for the diagonal slippage. Since phase only has meaning with respect to a cycle, it is possible to redraw Figure 8.13(a) to produce Figure 8.13(b). The timing of sync with respect to subcarrier now takes on the appearance of a frame rate sawtooth. The sync edge has a finite slope, and so several samples will be taken during the transition. Figure 8.13(c) shows how the phase of the instant when the 50% level of sync is crossed can be computed to a fraction of a subcarrier cycle. Since this position moves by a fixed amount per line (0.36 ns), it is possible to subtract a time which is obtained by multiplying the line number in the frame by 0.36 ns. This process is called horizontal modulation, and results in a modulated sync edge which appears to have four phases which are stationary with respect to subcarrier. The four phases are due to the quarter cycle offset. A further offset of one sample period is subtracted on alternate lines according to the state of 7.8 kHz, giving modulated sync edges which now have two phases. During the colour frame, the sync edge will be in-phase with subcarrier; during frame 3 it will be anti-phase. The colour frame is thus uniquely identified. If the input is not exactly EBU ScH phase, there will be a fixed phase error in the sync timing, and the system must pick the sync edge which is most nearly in phase with subcarrier to identify frame 1.

8.2.7 Digital colour framing in NTSC

The four-field sequence of NTSC arises because of the half cycle offset of subcarrier with respect to line frequency. The phase-locked loop which produces the sampling clock synchronizes to burst, but with a 57 degree offset so that samples are taken on the I and Q axes. The standard ScH phase relationship will occur on the first line of the four-field sequence, and on every other line. On the lines between, the sync edge will be located two sample periods away. These lines are easily detected and so the colour framing count is readily produced.

8.3 Digital video interface standards – summary

Standards or advanced proposals now exist for both 4:2:2 component and $4F_{sc}$ composite digital in both 525/59.94 and 625/50 and in both parallel and serial forms. There are also interfaces for 16:9 signals in both standard and high definition.

Digital interfaces require to be standardized in the following areas: connectors, to ensure plugs mate with sockets; pinouts; electrical signal specification, to ensure that the correct voltages and timing are transferred; and protocol, to ensure that the meaning of the data words conveyed is the same to both devices. As digital video of any type is only data, it follows that the same physical and electrical standards can be used for a variety of protocols.

The parallel interface uses common connectors, pinouts and electrical levels for both line standards in both the component and composite versions. The same is true for the serial interfaces. Thus there are only two electrical interfaces: parallel and serial. The type of video being transferred is taken care of in the protocol differences.

8.4 Parallel video interfaces

8.4.1 The parallel electrical interface

Composite digital signals use the same electrical and mechanical interface as is used for 4:2:2 component working[4,5]. This means that it is possible erroneously to plug a component signal into a composite machine. Whilst this cannot possibly work, no harm will be done because the signal levels and pinouts are the same.

A 25-pin D-type connector to ISO 2110-1989 is specified. Equipment always has female connectors, cables always have male connectors. Metal or metallized backshells are recommended with screened cables for optimum shielding. Equipment has been seen using ribbon cables and IDC (insulation displacement) connectors, but is it not clear whether such cables would meet the newer more stringent EMC (electromagnetic compatibility) regulations. It should be borne in mind that the ninth and eighteenth harmonics of 13.5 MHz are both emergency frequencies for aircraft radio.

Whilst equipment may produce or accept only 8 bit data, cables must contain conductors for all ten bits. Connector latching is by a pair of 4-40 (an American thread) screws, with suitable posts provided on the female connector. It is important that the screws are used as the multicore cable is quite stiff and can eventually unseat the plug if it is not secured. Some early equipment had slidelocks instead of screw pillars, but these proved to be too flimsy. During the changeover from slidelocks to 4-40 screws, some equipment was made with metric screw pillars and these will need to be changed to attach modern cables.

When unscrewing the locking screws from a D-connector it is advisable to check that the lock screw is actually unscrewing from the pillar. It is not unknown for the pillar to rotate instead. If this is not noticed, the pillar fixings may become detached inside the equipment, which will then have to be dismantled.

Each signal in the interface is carried by a balanced pair using ECL (emitter

Old 8-bit system	Connector pin number	New 10-bit system
Clock	1	Clock
System ground A	2	System ground A
Data 7 (MSB)	3	Data 9 (MSB)
Data 6	4	Data 8
Data 5	5	Data 7
Data 4	6	Data 6
Data 3	7	Data 5
Data 2	8	Data 4
Data 1	9	Data 3
Data 0	10	Data 2
Data A	11	Data 1
Data B	12	Data 0
Cable shield	13	Cable shield
Clock return	14	Clock return
System ground B	15	System ground B
Data 7 return	16	Data 9 return
Data 6 return	17	Data 8 return
Data 5 return	18	Data 7 return
Data 4 return	19	Data 6 return
Data 3 return	20	Data 5 return
Data 2 return	21	Data 4 return
Data 1 return	22	Data 3 return
Data 0 return	23	Data 2 return
Data A return	24	Data 1 return
Data B return	25	Data 0 return

(a)

(b)

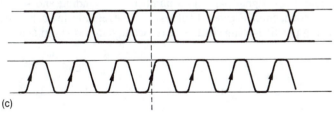

(c)

Figure 8.14 The parallel interface was originally specified as in (a) with eight bits expandable to ten. Later documents specify a 10 bit system (b) in which the bottom two bits may be unused. Clock timing at (c) is arranged so that a clock transition occurs between data transitions to allow maximum settling time

coupled logic) drive levels. The cable has a nominal impedance of 110 ohm and must be correctly terminated. ECL runs from a power supply of nominally –5.2 V and the logic states are –0.8 V for a 'high' and –1.85 V for a 'low'. ECL is primarily a current driven system, and the signal amplitude is quite low compared with other logic families as well as being negative valued.

Figure 8.14 shows the pinouts used. Although it is not obvious from the figure, the numbering of the D-connector is such that signal pairs are on physically opposite pins. Originally most equipment used 8 bit data and 10 bit working was viewed as an option; this was reflected in the wording of the first standards. However, in order to reflect the increasing quantity of 10 bit equipment now in use, the wording of later standards has subtly changed to describe a 10 bit system in which only eight bits may be used.

In the old specification shown in Figure 8.14(a), there are eight signal pairs and two optional pairs, so that extension to a 10 bit word can be accommodated. The optional signals were used to add bits at the least significant end of the word. Adding bits in this way extends resolution rather than increasing the magnitude. It will be seen from the figure that the optional bits are called Data –1 and Data –2 where the –1 and –2 refer to the powers of two represented, i.e. 2^{-1} and 2^{-2}. The 8 bit word describes 256 levels and ends in a radix point. The extra bits below the radix point represent the half and quarter quantizing intervals. In this way a degree of compatibility exists between 10 and 8 bit systems, as the correct magnitude will always be obtained when changing wordlength, and all that is lost is a degree of resolution in shortening the wordlength when the bits below the radix point are lost. The same numbering scheme can be used for both wordlengths; the longer wordlength simply has a radix point and an extra digit in any number base. Converting to the 8 bit equivalent can then be simply a matter of deleting the extra digit and retaining the integer part.

However, the later specification shown in Figure 8.14(b) renumbers the bits from 0 to 9 and assumes a system with 1024 levels. Thus all standard levels defined in the old 8 + 2 bit documents have to be multiplied by four to convert them to the levels in the new 10 bit documents. Thus a level of 16 decimal or 10 hex in the 8 bit system becomes 64 decimal or 40 hex in the 10 bit system. Figure 8.15 shows that 8 + 2 schemes may use hexadecimal numbering in the XYZ or XYLo format where two hex digits X and Y represent the most significant eight bits and the remaining two bits are represented by the Z or Lo symbol. 10 bit schemes are numbered with three hex digits PQR where P only has two meaningful bits.

A separate clock signal pair and a number of grounding and shielding pins complete the connection. Figure 8.14 also shows the relationship between the clock and the data. A positive-going clock edge is used to sample the signal lines after the

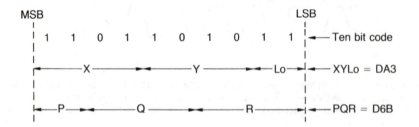

Figure 8.15 As there are two ways to parse ten bits into hexadecimal, there are two numbering schemes in use. In the XYLo system the parsing begins from the MSB down and the Lo parameter is expressed in quarters. In the PQR system the parsing starts from the LSB and so the P parameter has a maximum value of 3

level has settled between transitions. In component, the clock will be line-locked 27 MHz or 36 MHz irrespective of the line standard, whereas in composite digital the clock will be four times the frequency of PAL or NTSC subcarrier.

The parallel interface is suitable for distances of up to 50 metres (27 MHz) or 40 metres (36 MHz). Beyond this distance equalization is likely to be necessary and skew or differential delay between signals may become a problem. Equalization of such a large number of signals is not economically viable and for longer distances a serial interface is a better option.

8.4.2 The component parallel interface

It is not necessary to digitize analog sync pulses in component systems, since the only useful video data are those sampled during the active line. As the sampling rate is derived from sync, it is only necessary to standardize the size and position of a digital active line and all other parts of the video waveform can be recreated at a later time. The position is specified as a given number of sampling clock periods from

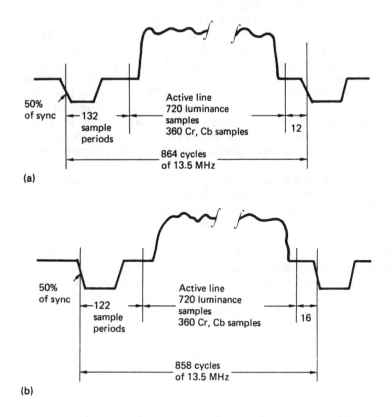

Figure 8.16 (a) In 625 line systems to CCIR-601, with 4:2:2 sampling, the sampling rate is exactly 864 times line rate, but only the active line is sampled, 132 sample periods after sync. (b) In 525 line systems to CCIR-601, with 4:2:2 sampling, the sampling rate is exactly 858 times line rate, but only the active line is sampled, 122 sample periods after sync. Note active line contains exactly the same quantity of data as for 50 Hz systems

the leading edge of sync, and the length is simply a standard number of samples. The component digital active line with 13.5 MHz sampling is 720 luminance samples long. This is somewhat longer than the analog active line and allows for some drift in the line position of the analog input. Ideally some of the first and last samples of the digital active line will represent blanking level, thereby avoiding abrupt signal transitions caused by a change from blanking level to active signal.

Figure 8.16 shows that in 625 line systems[5] the control system waits for 132 sample periods after an analog sync edge before commencing sampling the line. Then 720 luminance samples and 360 of each type of colour difference sample are taken; 1440 samples in all. A further 12 sample periods will elapse before the next sync edge, making $132 + 720 + 12 = 864$ sample periods. In 525 line systems[6], the analog active line is in a slightly different place and so the controller waits 122 sample periods before taking the same digital active line samples as before. There will then be 16 sample periods before the next sync edge, making $122 + 720 + 16 = 858$ sample periods.

For 16:9 aspect ratio working, the line and field rate remain the same, but the luminance sampling rate may be raised to 18 Mhz and the colour difference sampling rates are raised to 9 MHz. This results in the sampling structure shown for 625 lines in Figure 8.17(a) and for 525 lines in Figure 8.17(b). There are now 960 luminance pixels and 2×480 colour difference pixels per active line. The parallel interface remains the same except that the clock rate rises to 36 MHz.

When converting analog signals to digital it is important that the analog unblanked picture should be correctly positioned within the line. In this way the analog line will be symmetrically disposed within the digital active line. If this is the case, when converting the data back to the analog domain, no additional blanking will be needed as the blanking at the ends of the original analog line will be recreated from the data. The DAC can pass the whole of the digital active line for conversion and the result will be a correctly timed analog line with blanking edges in the right position.

However, if the original analog timing was incorrect, the unblanked analog line may be too long or off centre in the digital active line. In this case a DAC may apply digital blanking to the line data prior to conversion. Some equipment gives the user the choice of using blanking in the data or locally applied blanking prior to conversion.

In addition to specifying the location of the samples, it is also necessary to standardize the relationship between the absolute analog voltage of the waveform and the digital code value used to express it so that all machines will interpret the numerical data in the same way. These relationships are in the voltage domain and are independent of the scanning standard used.

Figure 8.18 shows how the luminance signal fits into the quantizing range of an 8 bit system. Numbering for 10 bit systems is shown with figures for eight bits in brackets. Black is at a level of 64_{10} (16_{10}) and peak white is at 940_{10} (235_{10}) so that there is some tolerance of imperfect analog signals and overshoots caused by filter ringing. The sync pulse will clearly go outside the quantizing range, but this is of no consequence as conventional syncs are not transmitted. The visible voltage range fills the quantizing range and this gives the best possible resolution.

The colour difference signals use offset binary, where 512_{10} (128_{10}) is the equivalent of blanking voltage. The peak analog limits are reached at 64_{10} (16_{10}) and 960_{10} (240_{10}) respectively, allowing once more some latitude for maladjusted analog inputs and filter ringing.

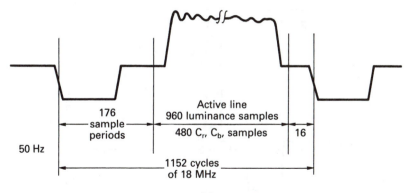

176
sample
periods

Active line
960 luminance samples

480 C$_r$, C$_b$, samples   16

50 Hz

1152 cycles
of 18 MHz

(a)

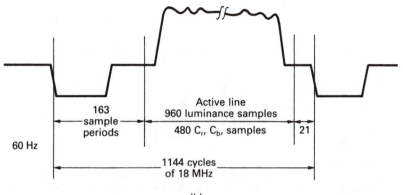

163
sample
periods

Active line
960 luminance samples

480 C$_r$, C$_b$, samples   21

60 Hz

1144 cycles
of 18 MHz

(b)

Figure 8.17 In 16:9 working with an 18 MHz sampling rate the sampling structure shown here results

Note that the code values corresponding to the eight most significant bits being all ones or all zeros (i.e. the two extreme ends of the quantizing range in 8 bit data) are not allowed to occur in the active line as they are reserved for synchronizing. ADCs must be followed by circuitry which catches these values and forces the LSB to a different value if out-of-range analog inputs are applied. Digital processing circuits which can generate these values must employ digital clamp circuits to remove the values from the signal. Fortunately this is a trivial operation.

The peak-to-peak amplitude of Y is 880 (220) quantizing intervals, whereas for the colour difference signals it is 900 (225) intervals. There is thus a small gain difference between the signals. This will be cancelled out by the opposing gain difference at any future DAC, but must be borne in mind when digitally converting to other standards.

The component interface carries a multiplex of luminance and colour difference samples and it is necessary to synchronize the demultiplexing process at the receiver so that the components are not inadvertently transposed. As conventional analog

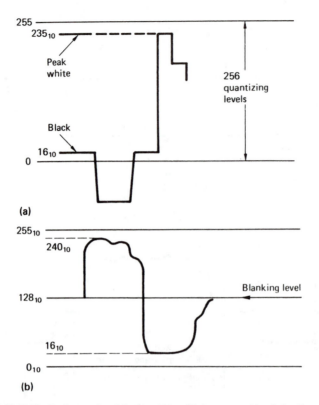

(a)

(b)

Figure 8.18 (a) The luminance signal fits into the quantizing range with a little allowance for excessive gain. Here black is at 16_{10} and peak white is at 235_{10}. The sync pulse goes outside the quantizing range but this is of no consequence as it is not transmitted. (b) The colour difference signals use offset binary where blanking level is at 128_{10}, and the peaks occur at 16_{10} and 240_{10} respectively

syncs are discarded, horizontal and vertical synchronizing must also be provided. These functions are performed by special bit patterns known as timing reference signals (TRS) sent with each line. Immediately before the digital active line location is the *SAV* (start of active video) TRS pattern, and immediately after is the *EAV* (end of active video) TRS pattern. These unique patterns occur on every line and continue throughout the vertical interval.

Each TRS pattern consists of four symbols; three of these form a sync pattern for demultiplexing and one is an information symbol which replaces the analog sync signals. As the parallel interface must work with 8 or 10 bit data, the sync pattern can only be specified for the eight most significant bits as the other two may not be present. The first symbol is eight ones and the next two are eight zeros. As these cannot occur in active video, their detection reliably indicates a sync pattern and is sufficient to enable unambiguous location of the information symbol and demultiplexing of the components. Ten bit sources should output ten ones and ten zeros during the TRS.

The fourth symbol is a data byte which contains three data bits, H, F and V. These bits are protected by four redundancy bits which form a 7 bit Hamming codeword.

Bit	9	8 F	7 V	6 H	5 P3	4 P2	3 P1	2 P0	1	0
	1	0	0	0	0	0	0	0	0	0
	1	0	0	1	1	1	0	1	0	0
	1	0	1	0	1	0	1	1	0	0
	1	0	1	1	0	1	1	0	0	0
	1	1	0	0	0	1	1	1	0	0
	1	1	0	1	1	0	1	0	0	0
	1	1	1	0	1	1	0	0	0	0
	1	1	1	1	0	0	0	1	0	0

Data   Check bits

(a)

Received P3 − P0	Received bits 8, 7, and 6 (F, V, and H)							
	000	001	010	011	100	101	110	111
0000	000	000	000	*	000	*	*	111
0001	000	*	*	111	*	111	111	111
0010	000	*	*	011	*	101	*	*
0011	*	*	010	*	100	*	*	111
0100	000	*	*	011	*	*	110	*
0101	*	001	*	*	100	*	*	111
0110	*	011	011	011	100	*	*	011
0111	100	*	*	011	100	100	100	*
1000	000	*	*	*	*	101	110	*
1001	*	001	010	*	*	*	*	111
1010	*	101	010	*	101	101	*	101
1011	010	*	010	010	*	101	010	*
1100	*	001	110	*	110	*	110	110
1101	001	001	*	001	*	001	010	*
1110	*	*	*	011	*	101	110	*
1111	*	001	010	*	100	*	*	*

(b)

Figure 8.19 The data bits in the TRS are protected with a Hamming code which is calculated according to the table in (a). Received errors are corrected according to the table in (b) where a dot shows an error detected but not correctable

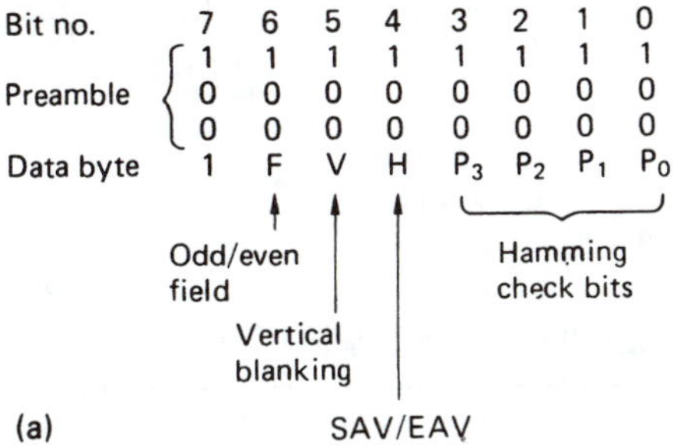

(a)   SAV/EAV

Figure 8.20(a) The 4 byte synchronizing pattern which precedes and follows every active line sample block has this structure

Figure 8.19(a) shows how the Hamming code is generated. Single bit errors can be corrected and double bit errors can be detected according to the decoding table in Figure 8.19(b).

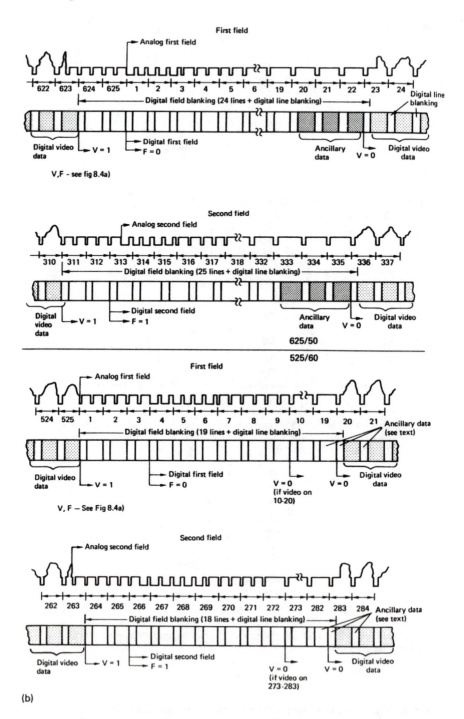

Figure 8.20(b) The relationships between analog video timing and the information in the digital timing reference signals for 625/50 (above) and 525/60 (below)

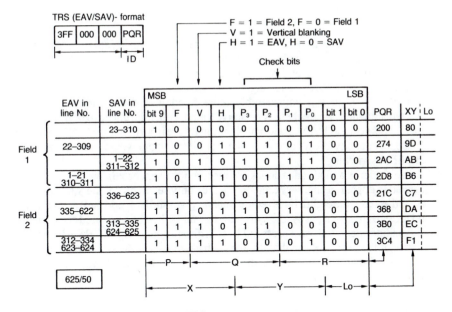

Figure 8.21 Decode table for component TRS

Figure 8.20(a) shows the structure of the TRS. The data bits have the following meanings:

- *H* is used to distinguish between *SAV*, where it is set to 0, and *EAV* where it is set to 1.

- *F* defines the state of interlace and is 0 during the first field and 1 during the second field. *F* is only allowed to change at *EAV*. In interlaced systems, one field begins at the centre of a line, but there is no sync pattern at that location so the field bit changes at the end of the line in which the change took place.

- *V* is 1 during vertical blanking and 0 during the active part of the field. It can only change at *EAV*.

Figure 8.20(b) (top) shows the relationship between the sync pattern bits and 625 line analog timing, whilst below is the relationship for 525 lines.

Figure 8.21 shows a decode table for TRS which is useful when interpreting logic analyser displays.

As only the active line is transmitted in component digital systems, DACs will need to have some additional circuitry to allow them to recreate analog syncs. As the sampling clock is horizontally locked only a small number of different sync edges need be stored and the memory requirements are minimal.

Note that since the syncs exceed the 8/10 bit digital signal range, and since they are only used for analog signals, a sync is sometimes regenerated in analog and added to the output of the DAC so as to avoid the requirement for 9/11 bit DACs.

8.4.3 Component ancillary data

Only the active line is transmitted and this leaves a good deal of spare capacity. The two line standards differ on how this capacity is used. In 625 lines, only the active line period may be used on lines 20 to 22 and 333 to 335[6]. Lines 20 and 333 are reserved for equipment self-testing.

In 525 lines there is considerably more freedom and ancillary data may be inserted anywhere where there is no active video, either during horizontal blanking where it is known as HANC, vertical blanking where it is known as VANC, or both[5]. The spare capacity allows many channels of digital audio which thus considerably simplifies switching.

The all zeros and all ones codes are reserved for synchronizing, and cannot be allowed to appear in ancillary data. In practice only seven bits of the 8 bit word can be used as data; the eighth bit is redundant and gives the byte odd parity. As all ones and all zeros are even parity, the sync pattern cannot then be generated accidentally.

Ancillary data is always prefaced by a different four-symbol TRS which is the inverse of the video TRS in that it starts with all zeros and then has two symbols of all ones followed by the information symbol. See section 8.7 for treatment of embedded audio in SDI.

8.4.4 The composite digital parallel interface

Composite digital samples at four times subcarrier frequency, and so there will be major differences between the PAL and NTSC standards. It is not possible to transmit digitized SECAM. Whilst the component interface transmits only active lines and special sync patterns, the composite interfaces carry the entire composite waveform – syncs, burst and active line. Although ancillary data may be placed in sync tip, the rising and falling sync edges must be present. In the absence of ancillary data, the data on the parallel interface is essentially the continuous stream of samples from a convertor which is digitizing a normal analog composite signal. Virtually all that is necessary to return to the analog domain is to strip out ancillary data and substitute sync tip values prior to driving a DAC and a filter. One of the reasons for this different approach is that the sampling clock in composite video is subcarrier locked. The sample values during sync can change with ScH phase in NTSC and PAL and change with the position in the frame in PAL due to the 25 Hz component. It is simpler to convey sync sample values on the interface than to go to the trouble of recreating them later.

The instantaneous voltage of composite video can go below blanking on dark saturated colours, and above peak white on bright colours. As a result the quantizing ranges need to be stretched in comparison with component in order to accommodate all possible voltage excursions. Sync tip can be accommodated at the low end and peak white is some way below the end of the scale. It is not so easy to determine when overload clipping will take place in composite as the sample sites are locked to subcarrier. The degree of clipping depends on the chroma phase. When samples are taken either side of a chroma peak, clipping will be less likely to occur than when the sample is taken at the peak. Advantage is taken of this phenomenon in PAL as the peak analog voltage of a 100% yellow bar goes outside the quantizing range. The sampling phase is such that samples are sited either side of the chroma peak and remain within the range.

The PAL and NTSC versions of the composite digital interface will be described separately. The electrical interface is the same for both, and was described in Section 8.4.1.

8.4.5 PAL interface

The quantizing range of digital PAL is shown in Figure 8.22[7]. Blanking level is at 256_{10} (64_{10}) and sync tip is the lowest allowable code of 4 (1) as 0 is reserved for digital synchronizing. Peak white is 844_{10} (211_{10}).

In PAL, the composite digital interface samples at $4 \times F_{sc}$ with sample phase aligned with burst phase. PAL burst swing results in burst phases of ±135 degrees, and samples are taken at these phases and at ±45 degrees, precisely half-way between the U and V axes. This sampling phase is easy to generate from burst and avoids premature clipping of chroma. It is most important that samples are taken exactly at the points specified, since any residual phase error in the sampling clock will cause the equivalent of a chroma phase error when samples from one source are added to samples from a different source in a switcher. A digital switcher can only add together pairs of samples from different inputs, but if these samples were not taken at the same instants with respect to their subcarriers, the samples represent different vectors and cannot be added.

Figure 8.23 shows how the sampling clock may be derived. The incoming sync is used to derive a burst gate, during which the samples of burst are analysed. If the clock is correctly phased, the sampled burst will give values of 380_{10} (95_{10}), 256_{10} (64_{10}), 128_{10} (32_{10}), 256_{10} (64_{10}), repeated, whereas if a phase error exists, the values at the burst crossings will be above or below 256_{10} (64_{10}). The difference between the sample values and blanking level can be used to drive a DAC which controls the

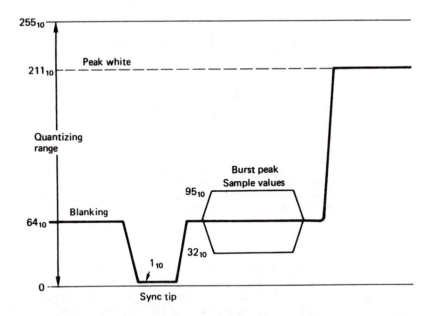

Figure 8.22 The composite PAL signal fits into the quantizing range as shown here. Note that there is sufficient range to allow the instantaneous voltage to exceed that of peak white in the presence of saturated bright colours. Values shown are decimal equivalents in a 10 or 8 bit system. In a 10 bit system the additional 2 bits increase resolution, not magnitude, so they are below the radix point and the decimal equivalent is unchanged. PAL samples in phase with burst, so the values shown are on the burst peaks and are thus also the values of the envelope

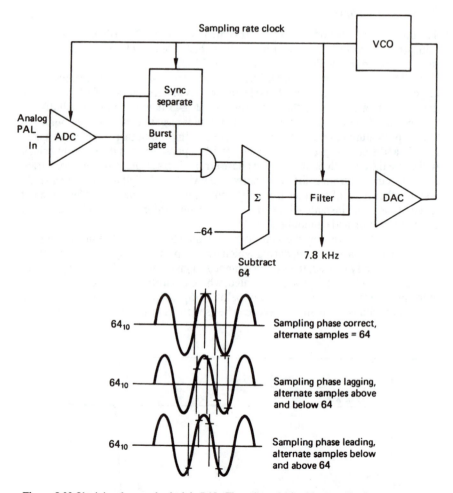

Figure 8.23 Obtaining the sample clock in PAL. The values obtained by sampling burst are analysed. When phase is correct, burst will be sampled at zero crossing and sample value will be 64_{10} or blanking level. If phase is wrong, sample will be above or below blanking. Filter must ignore alternate samples at burst peaks and shift one sample every line to allow for burst swing. It also averages over several burst crossings to reduce jitter. Filter output drives DAC and thus controls sampling clock VCO

sampling VCO. In this way any phase errors in the ADC are eliminated, because the sampling clock will automatically servo its phase to be identical to digital burst. Burst swing causes the burst peak and burst crossing samples to change places, so a phase comparison is always possible during burst. DC level shifts can be removed by using both positive and negative burst crossings and averaging the results. This also has the effect of reducing the effect of noise.

In PAL, the subcarrier frequency contains a 25 Hz offset, and so $4 \times F_{sc}$ will contain a 100 Hz offset. The sampling rate is not h-coherent, and the sampling structure is not quite orthogonal. As subcarrier is given by:

$$F_{sc} = 283.75 \times F_h + F_v/2$$

the sampling rate will be given by:

$$F_s = 1135F_h + 2F_v$$

This results in 709 379 samples per frame, and there will not be a whole number of samples in a line. In practice, 1135 sample periods, numbered 0 to 1134, are defined as one digital line, with an additional 2 sample periods per field which are included by having 1137 samples, numbered 0 to 1136, in lines 313 and 625. Figure 8.24(a) shows the sample numbering scheme for an entire line. Note that the sample numbering begins at 0 at the start of the digital active line so that the horizontal blanking area is near the end of the digital line and the sample numbers will be large. The digital active line is 948 samples long and is longer than the analog active line. This allows the digital active line to move with 25 Hz whilst ensuring the entire analog active line is still conveyed.

Since sampling is not h-coherent, the position of sync pulses will change relative to the sampling points from line to line. The relationship can also be changed by the

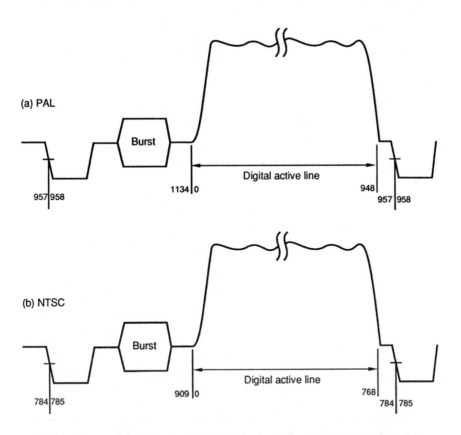

Figure 8.24 (a) Sample numbering in digital PAL. There are defined to be 1135 sample periods per line of which 948 are the digital active line. This is longer than the analog active line. Two lines per frame have two extra samples to compensate for the 25 Hz offset in subcarrier. NTSC is shown at (b). Here there are 910 samples per line of which 768 are the digital active line

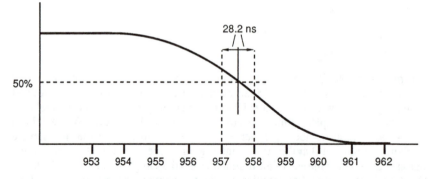

Figure 8.25 As PAL is sampled half-way between the colour axes, sample sites will fall either side of 50% sync at the zero ScH measurement line

ScH phase of the analog input. Zero ScH is defined as coincidence between sync and zero degrees of subcarrier phase at line 1 of field 1. Since composite digital samples on burst phase, not on subcarrier phase, the definition of zero ScH will be as shown in Figure 8.25, where it will be seen that two samples occur at exactly equal distances either side of the 50% sync point. If the input is not zero ScH, the samples conveying sync will have different values. Measurement of these values will allow ScH phase to be computed. In a DVTR installation, non-standard ScH is only a problem on the initial conversion from analog because composite DVTRs do not record sync, and will regenerate zero ScH syncs on replay. If non-zero ScH is input to a DVTR the output will be shifted horizontally on playback. Many composite digital VTRs are less tolerant of ScH phase variation than analog machines.

8.4.6 NTSC interface

Although they have some similarities, PAL and NTSC are quite different when analysed at the digital sample level. Figure 8.26 shows how the NTSC waveform fits into the quantizing structure[8]. Blanking is at 240_{10} (60_{10}) and peak white is at 800_{10} (200_{10}), so that 1 IRE unit is the equivalent of $1.4Q$ which could perhaps be called the DIRE. These different values are due to the different sync/vision ratio of NTSC. PAL is 7:3 whereas NTSC is 10:4.

Subcarrier in NTSC has an exact half-line offset, so there will be an integer number of cycles of subcarrier in two lines. F_{sc} is simply $227.5 \times F_h$, and as sampling is at $4 \times F_{sc}$, there will be $227.5 \times 4 = 910$ samples per line period, and the sampling will be orthogonal. Figure 8.24(b) shows that the digital active line consists of 768 samples numbered 0 to 767. Horizontal blanking follows the digital active line in sample numbers 768 to 909.

The sampling phase is chosen to facilitate encoding and decoding in the digital domain. In NTSC there is a phase shift of 123 degrees between subcarrier and the I axis. As burst is an inverted piece of the subcarrier waveform, there is a phase shift of 57 degrees between burst and the I axis. Composite digital NTSC does not sample in phase with burst, but on the I and Q axes at 57, 147, 237 and 327 degrees with respect to burst.

Figure 8.27 shows how this approach works in relation to sync and burst. Zero ScH is defined as zero degrees of subcarrier at the 50% point on sync, but the 57

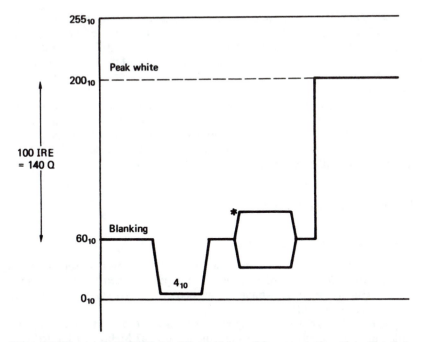

Figure 8.26 The composite NTSC signal fits into the quantizing range as shown here. Note that there is sufficient range to allow the instantaneous voltage to exceed peak white in the presence of saturated, bright colours. Values shown are decimal equivalents in an 8 or 10 bit system. In a 10 bit system the additional 2 bits increase resolution, not magnitude, so they are below the radix point and the decimal equivalent is unchanged. * Note that, unlike PAL, NTSC does not sample on burst phase and so values during burst are not shown here. See Figure 8.27 for burst sample details

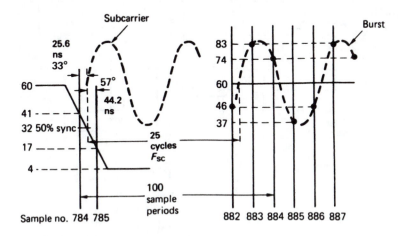

Figure 8.27 NTSC ScH phase. Sampling is not performed in phase with burst as in PAL, but on the *I* and *Q* axes. Since in NTSC there is a phase angle of 57° between burst and *I*, this will also be the phase at which burst samples should be taken. If ScH phase is zero, then phase of subcarrier taken at 50% sync will be zero, and the samples will be taken 33° before and 57° after sync; 25 cycles of subcarrier or 100 samples later, during burst, the sample values will be obtained. Note that in NTSC burst is inverted subcarrier, so sample 785 is positive, but sample 885 is negative

degree sampling phase means that the sync edge is actually sampled 25.6 ns ahead of, and 44.2 ns after, the 50% point. Similarly, when the burst is reached, the phase shift means that burst sample values will be 46_{10}, 83_{10}, 74_{10} and 37_{10} repeating. The phase-locked loop which produces the sampling clock will digitally compare the samples of burst with the values given here. Since it is not sampling burst at a zero crossing, the slope will be slightly less, so the gain of the phase error detector will also be less, and more prone to burst noise than in the PAL process. The phase error will normally be averaged over several burst samples to overcome this problem.

As in PAL, if the analog input does not have zero ScH phase, the sync pulse values will change, but burst values will not. As in PAL, NTSC DVTRs do not record sync, and will regenerate zero ScH syncs on replay.

8.5 Serial video interfaces

8.5.1 Component serial interface

The obsolescent CCIR 656[4,9] serial interface was designed to support 4:2:2 component data with 8 bit wordlength as, for example, recorded by the D-1 format. It has no 10 bit or composite variants. Group coding is used to allow transmission over 75 ohm coaxial cable.

Figure 8.28 shows the major components necessary. The coding used is the 8/9 code described in Chapter 3. Parallel 4:2:2 input data at 27 MHz according to CCIR 601/RP 125 having 8 bit wordlength are applied to the 8/9 convertor. This may be a PROM or gate array which is programmed to output a unique 9 bit pattern for each input data combination. These channel patterns are parallel loaded into a shift

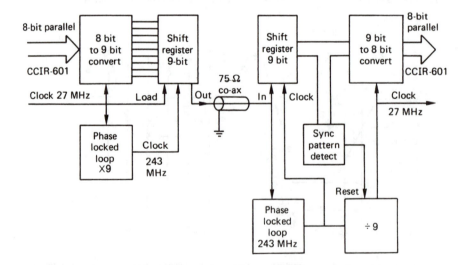

Figure 8.28 The major components necessary for a CCIR standard 8 bit serial link. Incoming data bytes are converted to 9 bit symbols which are clocked out of a shift register serially by a 243 MHz clock obtained from input 27 MHz. On reception a 243 MHz PLL locks to the data and produces a clock. When the sync pattern is seen, the divide-by-nine counter is re-phased to grab 9 bit symbols correctly from the shift register, which are then reconverted to 8 bits

register which is clocked at 243 MHz. Note that this is one of the aircraft emergency distress frequencies and circuits using this frequency need careful EMI screening. The clock is obtained by multiplying the input 27 MHz clock by 9 in a phase-locked loop. The serial signal from the shift register is sent to a driver stage and then to the cable.

The receiver has a number of tasks. Firstly it must lock to the channel bit rate and generate a local continuous 243 MHz clock from transitions in the serial waveform. The choice of the 9 bit codes is partly determined by their transition content. This is much higher than would be the case with serial raw data. The input waveform is sliced and sampled by the local clock to return it to a bit stream. Secondly the beginning of the 9 bit patterns must be identified so that the serial channel bit stream can be correctly deserialized. Only then can the 9 bit patterns be decoded to the original 8 bit data. This synchronizing step is made possible by the unique 9 bit sync patterns which result from coding the two all-zeros video data bytes which are present in SAV and EAV. The serialized 9 bit codes produce run lengths of 7 channel bits (011111110, 100000001) which cannot occur in any data symbol. The combination of the two codes is DC free. The receiver can detect this pattern by a combination of gates which inspect the parallel outputs of a shift register through which the input passes. When the sync pattern is detected, a reset can be applied to a divide-by-nine counter which runs from the received bit rate and clocks 9 bit parallel patterns from the shift register to the 9/8 decoder. This 27 MHz clock is also the parallel output word clock. The receiver may also check the validity of the Hamming coded sync data.

8.5.2 Serial digital interface (SDI)

The interface described here has been developed to allow up to 10 bit samples of component or composite digital video to be communicated serially[10]. The interface allows ancillary data including conveyance of AES/EBU digital audio channels.

Scrambling, or pseudo-random coding, uses concepts which were introduced in Chapter 3. The serial interface uses convolutional coding, which is simpler to implement in a cable installation because no separate synchronizing of the

System	Clock	Fundamental	Crash knee length	Practical length
NTSC Composite	143 MHz	71.5 MHz	400 m	320 m
PAL Composite	177 MHz	88.5 MHz	360 m	290 m
Component 601	270 MHz	135 MHz	290 m	230 m
Component 16:9	360 MHz	180 MHz	210 m	170 m
CABLE: BICC TM3205, PSF1/2, BELDEN 8281 or any cable with a loss of 8.7 dB/100 m at 100 MHz				

Figure 8.29 Suggested maximum cable lengths as a function of cable type and data rate to give a loss of no more than 30 dB. It is unwise to exceed these lengths due to the 'crash knee' characteristic of SDI

randomizing is needed. SDI operates with cable losses up to 30 dB. The losses increase with frequency and so the video standard in use and the grade of cable employed both affect the maximum distance which the signal will safely travel. Figure 8.29 gives some examples of cable lengths which can be used.

The components necessary for a composite serial link are shown in Figure 8.30. Parallel component or composite data having a wordlength of up to ten bits form the input. These are fed to a 10 bit shift register which is clocked at ten times the input rate, which will be 360 MHz, 270 MHz or $40 \times F_{sc}$. If there are only eight bits in the input words, the missing bits are forced to zero for transmission except for the all-ones condition which will be forced to ten ones. The serial data emerge from the shift register LSB first and are then passed through the scrambler, in which a given bit is converted to the exclusive OR of itself and two bits which are five and nine clocks ahead. This is followed by another stage, which converts channel ones into transitions. The resulting logic level signal is converted to a waveform which is symmetrical about ground and has an amplitude of 800 mV pk–pk across a 75 ohm load. This signal can be fed down 75 ohm coaxial cable using BNC connectors. SDI is restricted to point-to-point links. Serial receivers contain correct termination which is permanently present and passive loopthrough is not possible. No attempt should be made to drive more than one load using T-pieces as this will result in signal reflections which seriously compromise the data integrity.

The scrambling process at the transmitter spreads the signal spectrum and makes that spectrum reasonably constant and independent of the picture content. The transition encoder ensures that the signal is polarity independent. It is possible to assess the degree of equalization necessary by comparing the energy in a low frequency band with that in higher frequencies. The greater the disparity, the more equalization is needed. Thus fully automatic cable equalization at the receiver is easily achieved. The receiver must generate a bit clock at 360 MHz, 270 MHz or $40 \times F_{sc}$ from the input signal, and this clock drives the input sampler and slicer which converts the cable waveform back to serial binary. The local bit clock also drives a circuit which simply reverses the scrambling at the transmitter. The first stage returns transitions to ones, and the second stage is a mirror image of the encoder which reverses the exclusive OR calculation to output the original data. Since transmission is serial, it is necessary to obtain word synchronization, so that correct deserialization can take place. Such descrambling results in error extension, but this is not a practical problem since link error rates are practically zero.

In the component parallel input, the *SAV* and *EAV* TRS patterns are present and the all-ones and all-zeros bit patterns these contain can be detected in the shift register and used to teach the deserializer the location of word boundaries in the transmission.

In the composite parallel interface signal, there are no digital sync patterns and it is necessary to create an equivalent which can be inserted at the serializer and subsequently stripped out at a serial-to-parallel convertor. The timing reference and identification signal (TRS-ID) is added during blanking, and the receiver can detect the patterns which it contains. TRS-ID consists of five words which are inserted just after the leading edge of video sync. Figure 8.31(a) shows the location of TRS-ID at samples 967–971 in PAL and Figure 8.31(b) shows the location at samples 790–794 in NTSC.

Out of the five words in TRS-ID, the first four are for synchronizing, and consist of a single word of all ones, followed by three words of all zeros. Note that the composite TRS contains an extra word of zeros compared with the component TRS

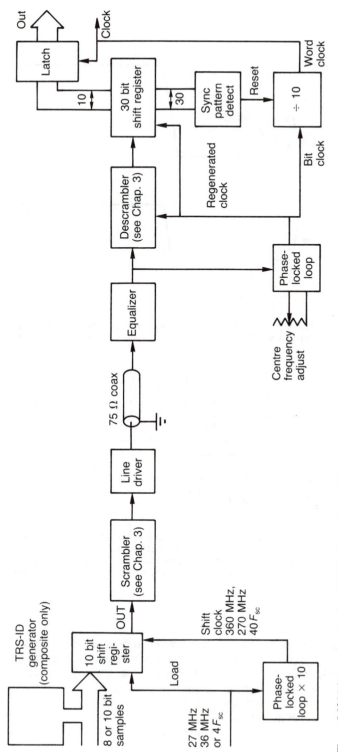

Figure 8.30 Major components of an SDI link. See text for details

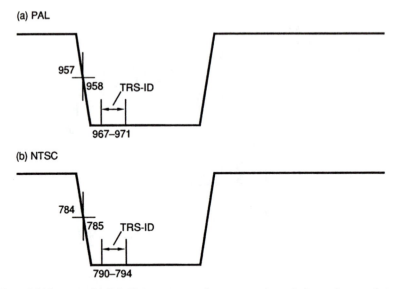

Figure 8.31 In composite digital it is necessary to insert a sync pattern during analog sync tip to ensure correct deserialization. The location of TRS-ID is shown at (a) for PAL and at (b) for NTSC

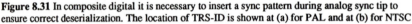

Figure 8.32 The contents of the TRS-ID pattern which is added to the transmission during the horizontal sync pulse just after the leading edge. The field number conveys the composite colour framing field count, and the line number carries a restricted line count intended to give vertical positioning information during the vertical interval. This count saturates at 31 for lines of that number and above

and this could be used for signal identification in multi-standard devices. The fifth word is for identification, and carries the line and field numbering information shown in Figure 8.32. The field numbering is colour framing information which is useful for editing. In PAL the field numbering will go from zero to seven, whereas in NTSC it will only reach three.

On detection of the synchronizing symbols, a divide-by-ten circuit is reset, and the output of this will clock words out of the shift register at the correct times. This circuit will also provide the output word clock.

8.6 Digital video interfacing chipsets

Implementation of digital video systems is much easier now that specialized chips are available. Figure 8.33 shows a hypothetical 4:2:2 component system starting with analog signals and ending with the same to illustrate the processes which are necessary. The syncs on Y are separated and multiplied in a phase-locked loop to produce a 27 MHz master clock. This is divided by 2 and by 4 to produce the sampling clocks for the convertors. This results in three data streams, which can be multiplexed to form a parallel interface signal using a parallel encoder chip such as the Sony CXD8068G. This parallel signal may be output using a set of ECL line drivers. If it is required to convert the parallel signal to SDI, a serial encoder will be required. The Sony SBX1610A and the Gennum GS9002 contain all parallel-to-serial functions, but output logic level signals which require a CXA 1389AQ or a GS9007 cable driver to produce the 1.6 volt pk–pk SDI signal which will fall to the standard 0.8 volts after passing through the source terminating resistors.

At the receiving end of the cable the signal requires equalization, clock regeneration and deserializing. The Sony SBX1602A provides all of these functions in one device whereas the Gennum solution is to combine equalization and reclocking in the GS9005 and to perform deserialization in the GS9000. In both cases the output is parallel single-ended data which can be returned to the parallel interface specification using ECL drivers. Alternatively the parallel data may be sent directly to a parallel interface decoder such as the Sony CXD8069G which demultiplexes the 27 MHz data to provide separate outputs for driving three DACs.

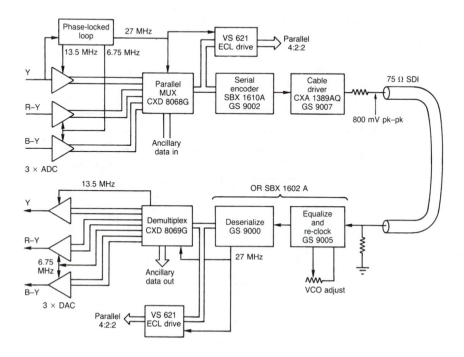

Figure 8.33 A hypothetical 4:2:2 system showing applications of various digital interfacing chips

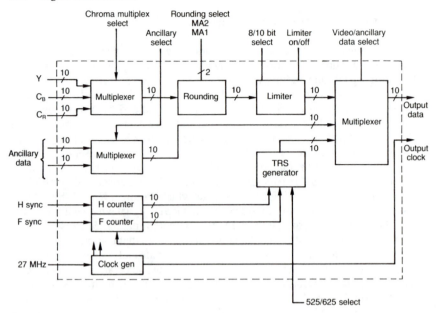

Figure 8.34 The three data streams from component ADCs can be multiplexed into the parallel interface format with the encoder chip shown here

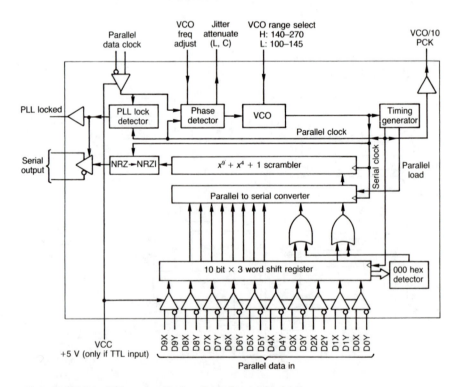

Figure 8.35(a) An SDI encoder chip from Sony. See text for details

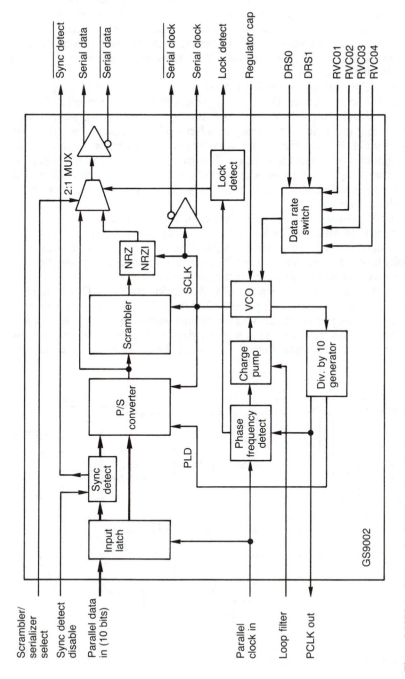

Figure 8.35(b) An SDI encoder chip from Gennum. See text for details

Figure 8.34 shows a block diagram of the CXD8068G parallel interface encoder. This accepts the parallel input from three component ADCs and multiplexes them to the 27 MHz parallel standard. The rounding process allows 10 bit inputs to be rounded to shorter wordlengths. The limiter prevents out-of-range analog signals from producing all-ones or all-zeros codes which are reserved for synchronizing. In addition to a 27 MHz clock derived from horizontal sync, the chip requires horizontal and frame drives to operate the timing counters which address the TRS generator. The final multiplexer selects TRS patterns, video data or ancillary data for the 10 bit parallel output.

Figure 8.35(a) shows the SBX1601A serial encoder and Figure 8.35(b) shows the GS9002 serial encoder. Of necessity these chips contain virtually identical processing. Parallel input data are clocked into the input latch by the parallel word clock which is multiplied in frequency by a factor of ten in a phase-locked loop to provide a serial bit clock. There is provision for selecting several centre frequencies for composite or component applications. The data latch output is examined by logic which detects input sync patterns and extends 8 bit sync values to ten bits. The parallel data are then serialized in a shift register prior to passing through the scrambler and the transition generator.

Figure 8.36 shows an SDI cable driver chip. The device shown has quadruple outputs and is useful in applications such as distribution amplifiers. Note that each differential amplifier produces a pair of separate SDI outputs. The fact that these are mutually inverted is irrelevant as the SDI signal is not polarity conscious. Note the resistor networks which provide correct cable source termination.

Figure 8.37(a) shows the Gennum GS9005 reclocking receiver. This contains an automatic cable equalizer and a phase-locked loop clock recovery circuit which drives a slicer/sampler which recovers the channel waveform to a logic level signal for subsequent descrambling in a separate device. The equalizer operates by estimating the cable length from the input amplitude and driving a voltage-

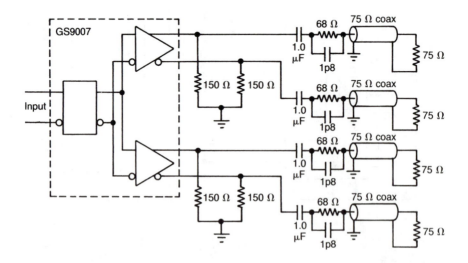

Figure 8.36 SDI cable driver chip provides correct 0.8 V signal level after source termination resistors

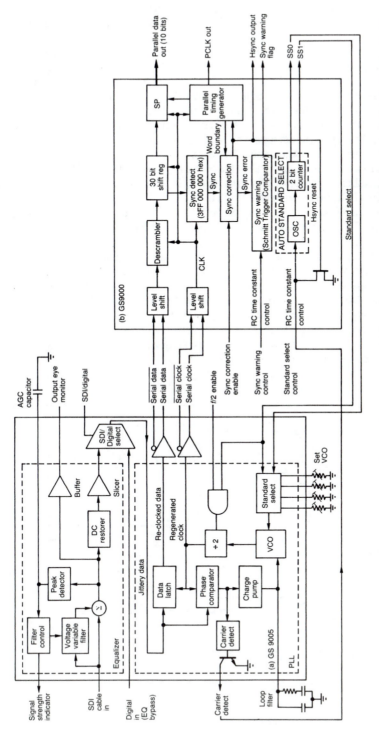

Figure 8.37 (a) Reclocking SDI receiver contains a cable equalizer and is an important building block for SDI routers as well as being the first stage of an SDI decoder. Decoder is shown in (b). Note auto standard sensing outputs which can select VCO frequency in the reclocker

controlled filter from the signal strength. A buffered eye pattern test point is provided. The equalizer output is DC restored prior to slicing to ensure that the slicing takes place around the waveform centre line. The slicer output will contain timing jitter and so a phase-locked loop is used having a loop filter to reject the jitter. The jitter-free clock is used to drive the data latch which samples the slicer output between transitions. The VCO centre frequency can be selected from four values and provision is made for an adjusting potentiometer for each frequency.

Figure 8.37(b) shows the GS9000 serial decoder which complements the GS9005. This contains a descrambler and a serial-to-parallel convertor which is synchronized by a sync detector which recognizes TRS in the shift register. The chip also contains an automatic standard detector which can output a 2 bit standard code for external indication and to select the centre frequency of the GS9005. The single-ended parallel output can be converted to the differential parallel output standard by a multiple ECL driver such as a VS621.

Figure 8.38 shows the Sony SBX1602A, which contains all of the serial receiving functions in one device. Its operation should be self-evident from the description of the Gennum devices above.

Parallel data can be demultiplexed for conversion to analog by the CXD8069G device shown in Figure 8.39 which also extracts ancillary data. The TRS detector identifies sync patterns and uses them to direct the ID word to the Hamming code error-correction stage. This outputs corrected timing signals which are decoded to produce analog video timing drives. A FIFO (first in first out) buffer acts as a small timebase corrector to allow the DACs to be driven with a stable clock. Ten-bit video data may be rounded to shorter wordlengths if required, prior to demultiplexing into separate component outputs.

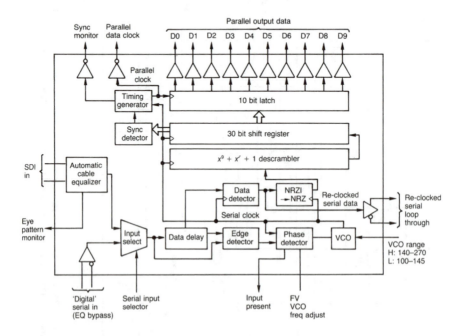

Figure 8.38 Sony SDI receiver chip for comparison with Figure 8.37

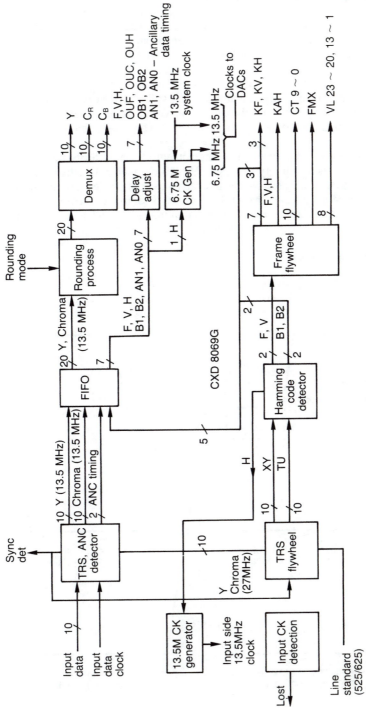

Figure 8.39 This device demultiplexes component data to drive separate DACs for each component as well as stripping out ancillary data

8.7 Embedded audio in SDI

In component SDI, there is provision for ancillary data packets to be sent during blanking[10,11]. The high clock rate of component means that there is capacity for up to sixteen audio channels sent in four groups. Composite SDI has to convey the digitized analog sync edges and bursts and only sync tip is available for ancillary data. As a result of this and the lower clock rate, composite has much less capacity for ancillary data than component although it is still possible to transmit one audio data packet carrying four audio channels in one group. Figure 8.40(a) shows where the ancillary data may be located for PAL and Figure 8.40(b) shows the locations for NTSC.

As was shown in Chapter 4, the data content of the AES/EBU digital audio subframe consists of validity (*V*), user (*U*) and channel (*C*) status bits, a 20 bit sample and four auxiliary bits which optionally may be appended to the main sample to produce a 24 bit sample. The AES recommends sampling rates of 48, 44.1 and 32 kHz, but the interface permits variable sampling rates. SDI has various levels of support for the wide range of audio possibilities and these levels are defined in

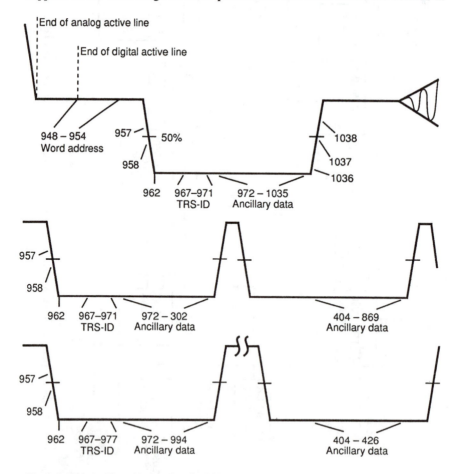

Figure 8.40(a) Ancillary data locations for PAL

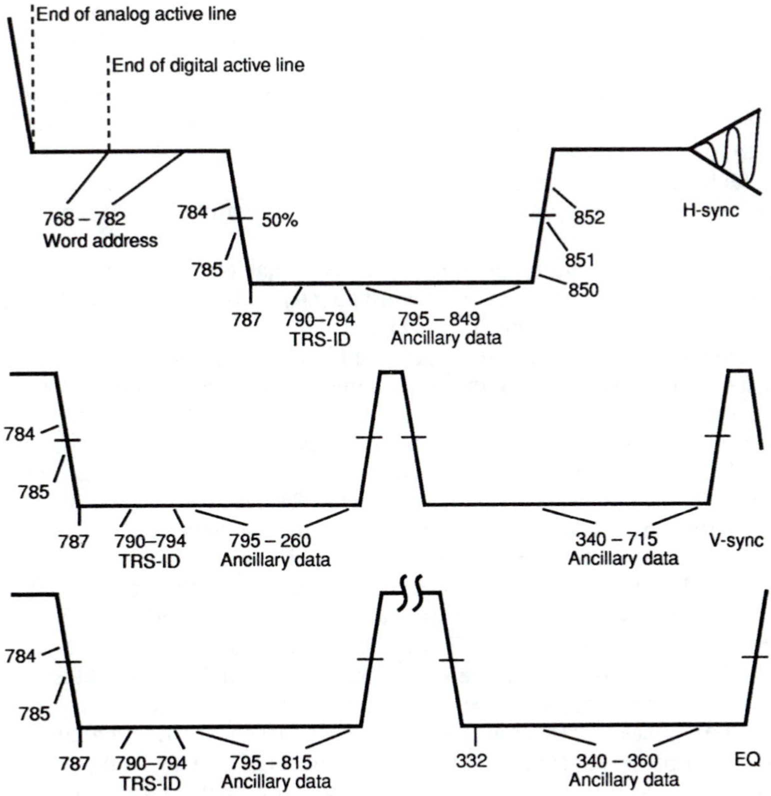

Figure 8.40(b) Ancillary data locations for NTSC

	A (Default)	Synchronous 48 kHz, 20 bit audio, 48 sample buffer
Needs audio control packet	B	Synchronous 48 kHz for composite video. 64 sample buffer to receive
		20 bits from 24 bit data
	C	Synchronous 48 kHz 24 bit with extended packets
	D	Asynchronous audio
	E	44.1 kHz audio
	F	32 kHz audio
	G	32–48 kHz variable sampling rate
	H	Audio frame sequence
	I	Time delay tracking
	J	Non-coincident channel status Z bits in a pair

Figure 8.41 The different levels of implementation of embedded audio. Level A is default

Figure 8.41. The default or minimum level is Level A which operates only with a video-synchronous 48 kHz sampling rate and transmits V, U, C and the main 20 bit sample only. As Level A is a default it need not be signalled to a receiver as the presence of IDs in the ancillary data is enough to ensure correct decoding. However, all other levels require an audio control packet to be transmitted to teach the receiver how to handle the embedded audio data. The audio control packet is transmitted once per field in the second horizontal ancillary space after the video switching point before any associated audio sample data. One audio control packet is required per group of audio channels.

If it is required to send 24 bit samples, the additional four bits of each sample are placed in extended data packets which must directly follow the associated group of audio samples in the same ancillary data space.

There are thus three kinds of packet used in embedded audio: the audio data packet which carries up to four channels of digital audio, the extended data packet and the audio control packet.

In component systems, ancillary data begin with a reversed TRS or sync pattern. Normal video receivers will not detect this pattern and so ancillary data cannot be mistaken for video samples. The ancillary data TRS consists of all zeros followed by all ones twice. There is no separate TRS for ancillary data in composite. Immediately following the usual TRS, there will be an ancillary data flag whose value must be $3FC_{16}$. Following the ancillary TRS or data flag is a data ID word which contains one of a number of standardized codes which tell the receiver how to interpret the ancillary packet. Figure 8.42 shows a list of ID codes for various types of packets. Next come the data block number and the data block count parameters. The data block number increments by 1 on each instance of a block with a given ID number. On reaching 255 it overflows and recommences counting. Next, a data count parameter specifies how many symbols of data are being sent in this block. Typical values for the data count are 36_{10} for a small packet and 48_{10} for a large packet. These parameters help an audio extractor to assemble contiguous data relating to a given set of audio channels.

Figure 8.43 shows the structure of the audio data packing. In order to prevent accidental generation of reserved synchronizing patterns, bit 9 is the inverse of bit 8 so the effective system wordlength is 9 bits. Three 9 bit symbols are used to convey all of the AES/EBU subframe data except for the four auxiliary bits. Since four audio channels can be conveyed, there are two 'Ch' or channel number bits which specify the audio channel number to which the subframe belongs. A further bit, Z, specifies the beginning of the 192 sample channel status message. V, U and C have the same significance as in the normal AES/EBU standard, but the P bit reflects parity on the three 9 bit symbols rather than the AES/EBU definition. The

	Group 1	Group 2	Group 3	Group 4
Audio data	2FF	1FD	1FB	2F9
Audio CTL	1EF	2EE	2ED	1EC
Ext. data	1FE	2FC	2FA	1F8

Figure 8.42 The different packet types have different ID codes as shown here

Address	x3	x3 + 1	x3 + 2
Bit			
B9	$\overline{\text{B8}}$	$\overline{\text{B8}}$	$\overline{\text{B8}}$
B8	A (2^5)	A (2^{14})	P
B7	A (2^4)	A (2^{13})	C
B6	A (2^3)	A (2^{12})	U
B5	A (2^2)	A (2^{11})	V
B4	A (2^1)	A (2^{10})	A MSB (2^{19})
B3	A LSB (2^0)	A (2^9)	A (2^{18})
B2	CH (MSB)	A (2^8)	A (2^{17})
B1	CH (LSB)	A (2^7)	A (2^{16})
B0	Z	A (2^6)	A (2^{15})

Figure 8.43 AES/EBU data for one audio sample is sent as three 9 bit symbols. A = audio sample. Bit Z = AES/EBU channel status block start bit

three-word sets representing an audio sample will then be repeated for the remaining three channels in the packet but with different combinations of the Ch bits.

One audio sample in each of the four channels of a group requires twelve video sample periods and so packets will contain multiples of twelve samples. At the end of each packet a checksum is calculated on the entire packet contents.

If 24 bit samples are required, extended data packets must be employed in which the additional four bits of each audio sample in an AES/EBU frame are assembled in pairs according to Figure 8.44. Thus for every twelve symbols conveying the four 20 bit audio samples of one group in an audio data packet two extra symbols will be required in an extended data packet.

The audio control packet structure is shown in Figure 8.45. Following the usual header are symbols representing the audio frame number, the sampling rate, the active channels, the processing delay and some reserved symbols. The sampling rate parameter allows the two AES/EBU channel pairs in a group to have different sampling rates if required. The active channel parameter simply describes which channels in a group carry meaningful audio data. The processing delay parameter denotes the delay the audio has experienced measured in audio sample periods. The parameter is a 26 bit two's complement number requiring three symbols for each channel. Since the four audio channels in a group are generally channel pairs, only two delay parameters are needed. However, if four independent channels are used, one parameter each will be required. The e bit denotes whether four individual channels or two pairs are being transmitted.

The frame number parameter comes about in 525 line systems because the frame rate is 29.97 Hz not 60 Hz. The resultant frame period does not contain a whole number of audio samples. An integer ratio is only obtained over a multiple frame sequence which is shown in Figure 8.46. The frame number conveys the position in the frame sequence. At 48 kHz odd frames hold 1602 samples and even frames hold 1601 samples in a five-frame sequence. At 44.1 and 32 kHz the relationship is not so simple and to obtain the correct number of samples in the sequence certain frames (exceptions) have the number of samples altered. At 44.1 kHz the frame sequence is 100 frames long whereas at 32 kHz it is 15 frames long. As the two channel pairs in a group can have different sampling rates, two frame parameters are required per group. In 50 Hz systems all three sampling rates allow an integer number of samples per frame and so the frame number is irrelevant.

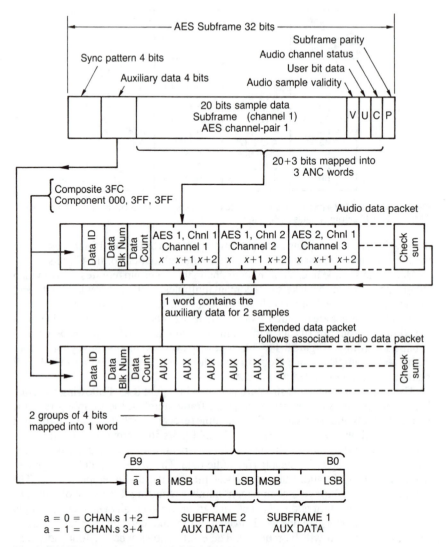

Figure 8.44 The structure of an extended data packet

As the ancillary data transfer is in bursts, it is necessary to provide a little RAM buffering at both ends of the link to allow real-time audio samples to be time compressed up to the video bit rate at the input and expanded back again at the receiver. Figure 8.47 shows a typical audio insertion unit in which the FIFO buffers can be seen. In such a system all that matters is that the average audio data rate is correct. Instantaneously there can be timing errors within the range of the buffers. Audio data cannot be embedded at the video switch point or in the areas reserved for EDH packets, but provided that data are evenly spread throughout the frame 20 bit audio can be embedded and retrieved with about 48 audio samples of buffering. If the additional four bits per sample are sent this requirement rises to 64 audio samples. The buffering stages cause the audio to be delayed with respect to

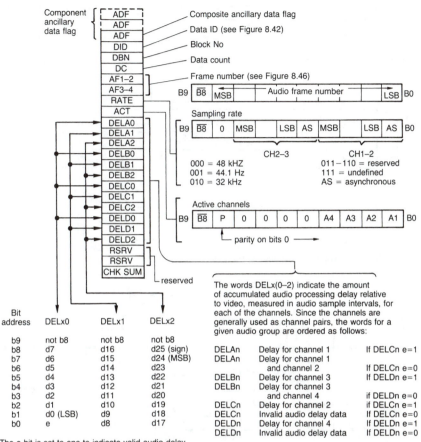

Bit address	DELx0	DELx1	DELx2
b9	not b8	not b8	not b8
b8	d7	d16	d25 (sign)
b7	d6	d15	d24 (MSB)
b6	d5	d14	d23
b5	d4	d13	d22
b4	d3	d12	d21
b3	d2	d11	d20
b2	d1	d10	d19
b1	d0 (LSB)	d9	d18
b0	e	d8	d17

The words DELx(0–2) indicate the amount of accumulated audio processing delay relative to video, measured in audio sample intervals, for each of the channels. Since the channels are generally used as channel pairs, the words for a given audio group are ordered as follows:

DELAn | Delay for channel 1 | If DELCn e=1
DELAn | Delay for channel 1 and channel 2 | If DELCn e=0
DELBn | Delay for channel 3 | If DELDn e=1
DELBn | Delay for channel 3 and channel 4 | if DELDn e=0
DELCn | Delay for channel 2 | if DELCn e=1
DELCn | Invalid audio delay data | If DELCn e=0
DELDn | Delay for channel 4 | If DELDn e=1
DELDn | Invalid audio delay data | If DELDn e=0

The e bit is set to one to indicate valid audio delay data. The delay words are referenced to the point where the AES/EBU data are input to the formatter. The delay words represent the average delay value, inherent in the formatting process, over a period no less than the length of the audio frame sequence (see Figure 8.46) plus any preexisting audio delay. Positive values indicate that the video leads the audio.

When only two channels are used, the e-bits in DELCn and DELDn must be set to 0 to indicate invalid while maintaining a constant size for the audio control packet.

The format for the audio delay is 26 bit two's complement.

Figure 8.45 The structure of an audio control packet

the video by a few milliseconds at each insertion. Whilst this is not serious, Level I allows a delay tracking mode which allows the embedding logic to transmit the encoding delay so a subsequent receiver can compute the overall delay. If the range of the buffering is exceeded for any reason, such as a non-synchronous audio sampling rate fed to a Level A encoder, audio samples are periodically skipped or repeated in order to bring the delay under control.

It is permitted for receivers which can only handle 20 bit audio to discard the 4 bit sample extension data. However, the presence of the extension data requires more buffering in the receiver. A device having a buffer of only 48 samples for Level A working could experience an overflow due to the presence of the extension data.

In 48 kHz working, the average number of audio samples per channel is just over

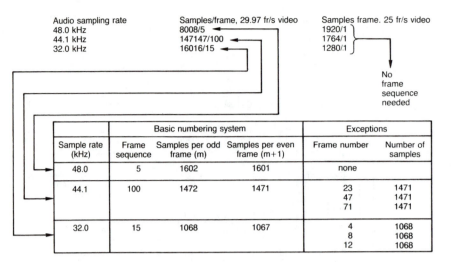

Figure 8.46 The origin of the frame sequences in 525 line systems

	Basic numbering system			Exceptions	
Sample rate (kHz)	Frame sequence	Samples per odd frame (m)	Samples per even frame (m+1)	Frame number	Number of samples
48.0	5	1602	1601	none	
44.1	100	1472	1471	23	1471
				47	1471
				71	1471
32.0	15	1068	1067	4	1068
				8	1068
				12	1068

Audio sampling rate	Samples/frame, 29.97 fr/s video	Samples frame. 25 fr/s video
48.0 kHz	8008/5	1920/1
44.1 kHz	147147/100	1764/1
32.0 kHz	16016/15	1280/1

No frame sequence needed

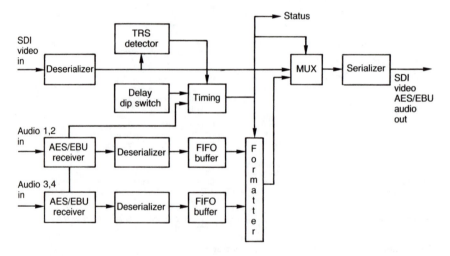

Figure 8.47 A typical audio insertion unit. See text for details

three per video line. In order to maintain the correct average audio sampling rate, the number of samples sent per line is variable and not specified in the standard. In practice a transmitter generally switches between packets containing three samples and packets containing four samples per channel per line as required to keep the buffers from overflowing. At lower sampling rates either smaller packets can be sent or packets can be omitted from certain lines.

As a result of the switching, ancillary data packets in component video occur mostly in two sizes. The larger packet is 55 words in length of which 48 words are data. The smaller packet contains 43 words of which 36 are data. There is space for two large packets or three small packets in the horizontal blanking between *EAV* and *SAV*.

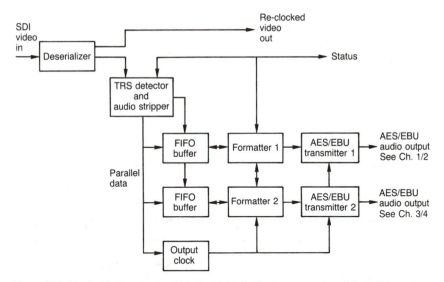

Figure 8.48 A typical audio extractor. Note the FIFOs for timebase expansion of the audio samples

A typical embedded audio extractor is shown in Figure 8.48. The extractor recognizes the ancillary data TRS or flag and then decodes the ID to determine the content of the packet. The group and channel addresses are then used to direct extracted symbols to the appropriate audio channel. A FIFO memory is used to timebase expand the symbols to the correct audio sampling rate.

8.8 EDH – error detection and handling

The serial digital interface is steadily taking over as the video interface standard for the digital age, but the original standard had no provisions for data integrity checking. EDH is an option for serial digital which goes a long way to rectifying the problem[12,13]. Figure 8.49 shows an EDH equipped SDI (serial digital interface) transmission system. At the first transmitter, the data from one field is transmitted and simultaneously fed to a cyclic redundancy check (CRC) generator. The CRC calculation is a mathematical division by a polynomial and the result is the remainder. The remainder is transmitted in a special ancillary data packet which is sent early during the vertical interval, before any switching takes place in a router[14]. The first receiver has an identical CRC generator which performs a calculation on the received field. The ancillary data extractor identifies the EDH packet and demultiplexes it from the main data stream. The remainder from the ancillary packet is then compared with the locally calculated remainder. If the transmission is error free, the two values will be identical. In this case no further action results. However, if as little as one bit is in error in the data, the remainders will not match. The remainder is a 16 bit word and guarantees to detect up to 16 bits in error anywhere in the field. Greater numbers of errors are not guaranteed to be detected, but this is of little consequence as enough fields in error will be detected to indicate that there is a problem.

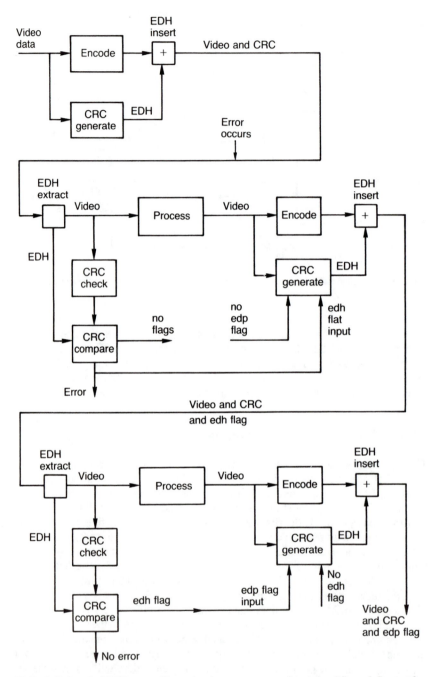

Figure 8.49 A typical EDH system illustrating the way errors are detected and flagged. See text for details

Should a CRC mismatch indicate an error in this way, two things happen. Firstly an optically isolated output connector on the receiving equipment will present a low impedance for a period of 1 to 2 milliseconds. This will result in a pulse in an externally powered circuit to indicate that a field contained an error. An external error-monitoring system wired to this connector can note the occurrence in a log or sound an alarm or whatever it is programmed to do. As the data is incorrectly received, the fact must also be conveyed to subsequent equipment. It is not permissible to pass on a mismatched remainder. The centre unit in Figure 8.49 must pass on the data as received, complete with errors, but it must calculate a new CRC which matches the erroneous data. When received by the third unit in Figure 8.49, there will then only be a CRC mismatch if the transmission between the second and third devices is in error. This is correct as the job of the CRC is to only to locate faulty hardware and clearly if the second link is not faulty the CRC comparison should not

Data item	b9 msb	b8	b7	b6	b5	b4	b3	b2	b1	b0 lsb
Ancillary data header, word 1 – component	0	0	0	0	0	0	0	0	0	0
Ancillary data header, word 2 – component	1	1	1	1	1	1	1	1	1	1
Ancillary data header, word 3 – component	1	1	1	1	1	1	1	1	1	1
Auxiliary data flag – composite	1	1	1	1	1	1	1	1	0	0
Data ID (1F4)	0	1	1	1	1	1	0	1	0	0
Block number	1	0	0	0	0	0	0	0	0	0
Data count	0	1	0	0	0	1	0	0	0	0
Active picture data word 0 crc<5:0>	$\bar{P}$	P	C_5	C_4	C_3	C_2	C_1	C_0	0	0
Active picture data word 1 crc<11:6>	$\bar{P}$	P	C_{11}	C_{10}	C_9	C_8	C_7	C_6	0	0
Active picture data word 2 crc<15:12>	$\bar{P}$	P	V	0	C_{15}	C_{14}	C_{13}	C_{12}	0	0
Full-field data word 0 crc<5:0>	$\bar{P}$	P	C_5	C_4	C_3	C_2	C_1	C_0	0	0
Full-field data word 1 crc<11:6>	$\bar{P}$	P	C_{11}	C_{10}	C_9	C_8	C_7	C_6	0	0
Full-field data word 2 crc<15:12>	$\bar{P}$	P	V	0	C_{15}	C_{14}	C_{13}	C_{12}	0	0
Auxiliary data error flags	$\bar{P}$	P	0	ues	ida	idh	eda	edh	0	0
Active picture error flags	$\bar{P}$	P	0	ues	ida	idh	eda	edh	0	0
Full-field error flags	$\bar{P}$	P	0	ues	ida	idh	eda	edh	0	0
Reserved words (7 total)	1	0	0	0	0	0	0	0	0	0
Checksum	$\bar{S8}$	S8	S7	S6	S5	S4	S3	S2	S1	S0

Error flags

All error flags indicate only the status of the previous field; that is, each flag is set or cleared on a field-by-field basis. A logical 1 is the set state and a logical 0 is the unset state. The flags are defined as follows:

edh – error detected here: Signifies that a serial transmission data error was detected. In the case of ancillary data, this means that one or more ANC data blocks did not match its checksum.

eda – error detected already: Signifies that a serial transmission data error has been detected somewhere upstream. If device B receives a signal from device A and device A has set the edh flag, when device B retransmits the data to device C, the eda flag will be set and the edg flag will be unset if there is no further error in the data.

idh – internal error detected here: Signifies that a hardware error unrelated to serial transmission has been detected within a device. This is provided specifically for devices which have internal data error checking facilities, as an error reporting mechanism.

ida – internal error detected already: Signifies that an idh flag was received and there was a hardware device failure somewhere upstream.

ues – unknown error status: Signifies that a serial signal was received from equipment not supporting this error-detection mechanism.

Checkword values

Each checkword value consists of 16 bits of data calculated using the CRC-CCITT polynomial generation method. The equation and a conceptual logic diagram are shown below:

Checkword (16-bit) $= x^{16} + x^{12} + x^5 + 1$

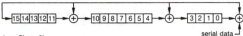

(see Chap. 3)

Figure 8.50 The contents of the EDH packet which is inserted in ancillary data space after the associated field

fail. However, the third device still needs to know that there is a problem with the data, and this is the job of the error flags which also reside in the EDH packet. One of these flags is called edh (error detected here) and this will be asserted by the centre device in Figure 8.49. The last device in Figure 8.49 will receive edh and transmit eda (error detected already). There are also flags to handle hardware failures (e.g. over temperature or diagnostic failure). The idh (internal error detected here) and ida (internal error detected already) handle this function. Locally detected hardware errors drive the error output socket to a low impedance state constantly to distinguish from the pulsing of a CRC mismatch.

A slight extra complexity is that error checking can be performed in two separate ways. One CRC is calculated for the active picture only, and another is calculated for the full field. Both are included in the EDH packet which is shown in Figure 8.50. The advantage of this arrangement is that whilst regular program material is being passed in active picture, test patterns can be sent in vertical blanking which can be monitored separately. Thus if active picture is received without error but full field gives an error, the error must be outside the picture. It is then possible to send, for example, pathological test patterns during the vertical interval which stress the transmission system more than regular data to check the performance margin of the system. This can be done alongside the picture information without causing any problems.

In a large system, if every SDI link is equipped with EDH, it is possible for automatic error location to be performed. Each EDH-equipped receiver is connected to a monitoring system which can graphically display on a map of the system the location of any transmission errors. If a suitable logging system is used, it is not necessary for the display to be in the same place as the equipment. In the event of an error condition, the logging system can communicate with the display by dial-up modem or dedicated line over any distance. Logging allows infrequent errors to be counted. Any increase in error rate indicates a potential failure which can be rectified before it becomes serious.

An increasing amount of new equipment is available with EDH circuitry. However, older equipment can still be incorporated into EDH systems by connecting it in series with proprietary EDH insertion and checking modules.

References

1   SMPTE 170M, Television Studio – System M/NTSC – Composite Analog Video Signal
2   CCIR Report 624-3, Characteristics of Television Systems
3   CCIR Recommendation 601-1, Encoding Parameters for Digital Television for Studios
4   CCIR Recommendation 656
5   SMPTE 125M, Television – Bit Parallel Digital Interface – Component Video Signal 4:2:2
6   EBU Doc. Tech. 3246
7   SMPTE Proposed Standard – Bit parallel digital interface for 625/50 system PAL composite digital video signals
8   SMPTE 244M, Television – System M/NTSC Composite Video Signals – Bit-Parallel Digital Interface
9   EBU Doc. Tech. 3247
10 SMPTE 259M – 10-bit 4:2:2 Component and $4F_{sc}$ NTSC Composite Digital Signals – Serial Digital Interface
11 WILKINSON, J. H. (1991) Digital audio in the digital video studio – a disappearing act? Presented at *9th International AES Conference, Detroit*, Audio Engineering Society

12 ELKIND, R. and FIBUSH, D. (1991) Proposal for error detection and handling in studio equipment. Presented at *25th SMPTE Television Conference, Detroit*

13 SMPTE RP165 – Error detection checkwords and status flags for use in bit-serial digital interfaces for television

14 SMPTE RP168 – Definition of vertical interval switching point for synchronous video switching

Chapter 9

Practical video interfacing

Having covered a lot of theoretical ground and described the detail of a number of standards earlier in this book, the purpose of this chapter is to give practical advice in configuring, timing, testing and maintaining digital video interface systems. The techniques and tools needed to test digital interfaces are quite different to the traditional analog approach and will be described here.

9.1 Digital video routing

Digital routers have the advantage that they need cause no loss of signal quality as they simply pass on a series of numbers. Analog routers inevitably suffer from crosstalk and noise, however well made, and this reduces signal quality on every pass.

Routers are available for serial and parallel formats. The parallel video output signal is cheaper to produce as most equipment has parallel format internally. On the other hand a parallel router is complex because of the large number of conductors which need switching for each input. Parallel digital video is also limited to around 50 metres of cable length and so cannot replace the traditional analog router.

A serial router is potentially very inexpensive as it is a single-pole device. It can be easier to build than an analog router because the digital signal is more resistant to crosstalk. Now that SDI chips are available, parallel working is becoming obsolete. Large routers can easily be implemented with SDI and the long cable drive capability means that large broadcast installations can be tackled.

A serial router can be made using wideband analog switches, so that the input waveform is passed from input to output. This is an inferior approach, as the total length of cable which can be used is restricted; the input cable is effectively in series with the output cable and the analog losses in both will add.

The correct approach is for each router input to re-clock and slice the waveform back to binary. Figure 9.1 shows a typical re-clocking SDI router. Following the input re-clocking stage the actual routing takes place on logic level signals prior to a line driver which re-launches a clean signal. The cables to and from the router can be maximum length as the router is effectively a repeater.

It is not necessary to unscramble the serial signal at the router. A phase-locked loop is used to regenerate the bit clock. This rejects jitter on the incoming waveform. The waveform is sliced, and the slicer output is sampled by the local clock. The result is a clean binary waveform, identical to the original driver waveform. The

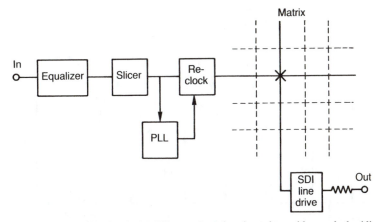

Figure 9.1 A router for SDI signals should be constructed as shown here with a re-clocker/slicer at each input to restore the received waveform to clean binary logic levels. The routing matrix proper is then a logic element which is followed by SDI line drivers. If this is not done the output cable is electrically in series with the input cable and the performance margin will be impaired. There is no need to descramble or decode the signal as the router is not interested in its meaning

Gennum GS9005 is an equalizer/re-clocker chip which is suitable for router applications as it does not descramble the data stream.

The router is simply a binary bit stream switch, and is not unduly concerned with the meaning or content of the bit stream. It does not matter whether the bit stream is PAL, NTSC or 4:2:2 or whether or not ancillary data is carried – the information just passes through.

The only parameter of any consequence is the bit rate. Component runs at 270 or 360 Mbits/sec, PAL runs at 177 Mbits/sec and NTSC runs at 143 Mbits/sec. Most multi-standard routers require a link or DIP switch to be set in each input in order to select the appropriate VCO centre frequency. A separate VCO adjustment will be present for each standard. Otherwise units can be standards independent, which allows more flexibility and economy. With a mixed standard router, it is only necessary to constrain the control software so that inputs of a given standard can only be routed to outputs connected to devices of the same standard and one router can then handle component and composite signals simultaneously.

9.2 Timing in digital installations

The issue of signal timing has always been critical in analog video, but the adoption of digital routing relaxes the requirements considerably. Analog vision mixers need to be fed by equal length cables from the router to prevent propagation delay variation. In the digital domain this is no longer an issue as delay is easily obtained and each input of a digital vision mixer can have its own local timebase corrector. Provided signals are received having timing within the window of the inputs, all inputs are re-timed to the same phase within the mixer.

Figure 9.2 shows how a mixing suite can be timed to a large SDI router. Signals to the router are phased so that the router output is aligned to station reference within

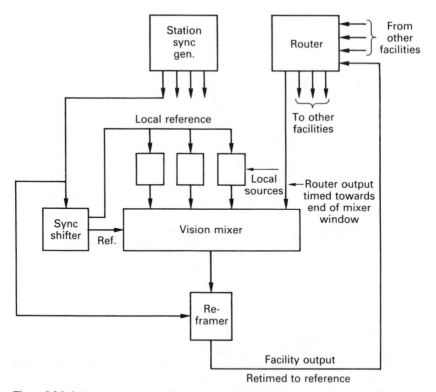

Figure 9.2 In large systems some care is necessary with signal timing as is shown here. See text for details

a microsecond or so. The delay in the router may vary with its configuration but only by a few microseconds. The mixer reference is set with respect to station reference so that local signals arrive towards the beginning of the input windows and signals from the router (which, having come further, will be the latest) arrive towards the end of the windows. Thus all sources can be re-timed within the mixer and any signal can be mixed with any other. Clearly the mixer introduces delay, and the signal feed back to the router experiences further delay. In order to send the mix back to the router a frame synchronizer is needed on the output of the suite. This introduces somewhat less than a frame of delay so that by the time the signal has re-emerged from the router it is aligned to station reference once more, but a frame late. An installation of this kind relies on a gen-lockable sync pulse generator having multiple outputs with independent phase control.

In an ideal world, every piece of hardware in the station would have component SDI outputs and inputs, and everything would be connected by SDI cables. In practice, unless the building is new, this is unlikely. However, there are ways in which SDI can be phased in alongside analog systems. An expensive way of doing this is to fit every composite device with coders and decoders, and every analog device with convertors so that a component SDI router can be used alone. Figure 9.3 shows an alternative. The SDI router is connected to every piece of digital equipment, using parallel-to-serial adapters where necessary. Analog VTRs such as

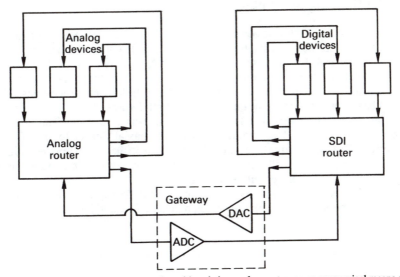

Figure 9.3 Digital routers can operate alongside existing analog routers as an economical means of introducing digital routing

Betacam SP or M-II have digital output options which use the data from the TBC directly. One aspect to check with SP and M-II machines having digital outputs is that the Y/C timing must be adjusted correctly at the TBC input in the VTR as it cannot be corrected subsequently. These will need DACs to convert SDI signals to analog component inputs. Analog equipment continues to be connected to an analog router, and interconnection paths, called gateways, are created between the two routers. These gateways require convertors in each direction, but the number of convertors is much less than if every analog device was equipped. The number of gateways will be determined by the number of simultaneous transactions between analog and digital domains.

In many cases the two routers can be made to appear like one large router if appropriate software is available in a common control system. This will only be possible if the SDI router is purchased from the same manufacturer as the existing analog router or if the SDI router manufacturer offers custom software.

Whilst modern DVTRs incorporate the audio data in the SDI signal, adapting older devices to do this and subsequently demultiplexing the audio could prove expensive. DVTRs tend also to be inflexible in their audio selection. For example it may not be possible to record certain audio channels from embedded SDI data at the same time as other channels from the AES/EBU inputs. Also the embedded audio may not contain timecode even though the standard allows it. As a result in some installations it may be more appropriate to retain an earlier audio router, or to have a separate AES/EBU digital audio router layer controlled in parallel with the SDI router.

If the analog router handles composite signals, then coders and decoders will be needed in addition to the convertors. In this case the use of gateways obviates unnecessary codecs when analog devices are connected together.

9.3 Configuring SDI links

The viability of an SDI link is governed primarily by the data rate and the proposed distance. It is quite easy to establish what grade of cable will be required from the tables in Chapter 8. As can be seen from Figure 9.4, the data integrity deteriorates rapidly beyond a critical cable length. The sudden upswing in bit error rate is known as the 'crash knee' and for reliability only operation to the left of the knee is possible. Do not be tempted to stretch the quoted cable length figures. As SDI is a point-to-point interface all receivers are equipped with an internal terminator. Thus there is no such thing as passive loop-through and use of a coaxial T-piece for monitoring is ruled out. The use of active loop-through means that, in the case of a power failure, the loop-through signal will fail. If it is required to drive multiple destinations, a digital distribution amplifier will be needed.

If the distance required is excessive even for the best grade of cable then a repeater will be required. Unlike an analog repeater, a properly engineered digital

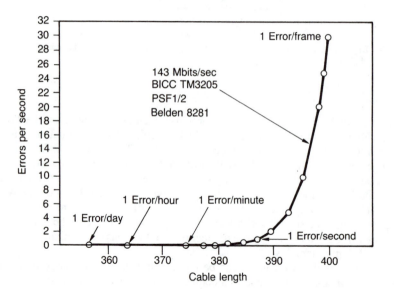

Bit error rates (BER)	NTSC	PAL	270 Mb	360 Mb
1 bit error/field	4.2×10^{-7}	2.8×10^{-7}	1.8×10^{-7}	1.3×10^{-7}
1 bit error/second	7.0×10^{-9}	4.7×10^{-9}	3.1×10^{-9}	2.3×10^{-9}
1 bit error/minute	1.2×10^{-10}	7.8×10^{-11}	5.1×10^{-11}	3.8×10^{-11}
1 bit error/hour	1.9×10^{-12}	1.3×10^{-12}	9.0×10^{-13}	6.4×10^{-13}
1 bit error/day	8.1×10^{-14}	5.4×10^{-14}	3.5×10^{-14}	2.7×10^{-14}
1 bit error/month	2.6×10^{-15}	1.8×10^{-15}	1.2×10^{-15}	8.9×10^{-16}
1 bit error/year	2.2×10^{-16}	1.5×10^{-16}	1.0×10^{-16}	7.4×10^{-17}

Figure 9.4 Because of the multiplicative effect of the large number of factors causing signal degradation the error rate increases steeply after a certain cable length. This sudden onset of errors is referred to as the 'crash knee'

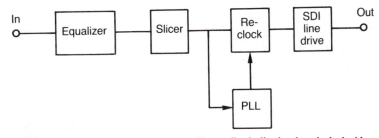

Figure 9.5 In a re-clocking repeater the input signal is equalized, sliced and re-clocked with a phase-locked loop to recover a clean binary signal which then passes to an SDI line driver

repeater causes no generation loss and minimal delay. Figure 9.5 shows a typical re-clocking repeater. The repeater is not interested in the meaning of the data and need not descramble or deserialize the data stream. It is only necessary to re-clock with a phase-locked loop to reject jitter and slice to reject noise. The resultant clean logic level signal can then be used to drive a further SDI waveform generator.

Such a simple re-clocking repeater is of limited use in an EDH system as there is no ability to distinguish errors occurring before or after the repeater. If this is important an EDH supporting repeater will be necessary. By definition such a device must descramble and de-serialize the data stream in order to make the error-detecting checks on the incoming data. As a result it will cause a greater signal delay than a simple re-clocking repeater.

9.4 Testing digital video interfaces

Once video and audio are converted to the digital domain, they become data, or numbers, and if those numbers can be delivered to the other end of a digital interface unchanged then the interface has not caused any loss of quality. This is one of the strengths of digital technology. In the absence of data reduction techniques, the quality is determined in the conversion process and can then be maintained in transmission and recording. In contrast analog signals are subject to generation loss in every recording and to noise and distortion in every transmission. This analog heritage has led to a philosophy where the analog waveform is monitored at every stage so that some adjustment can be made to minimize the quality loss. The waveform monitor and vectorscope tradition is so strong that despite the transition to the radically different digital technology many people think no new monitoring methods are needed.

Unfortunately, traditional analog testing techniques reveal nothing about a digital interface or recorder. Consider the system of Figure 9.6, which could be composite or component. An ADC converts the input waveform to data which is transmitted by the interface. A DAC converts the received data to analog video once more. If a waveform monitor and vectorscope are connected to the DAC, what information is revealed about the interface? If it is assumed that the interface is not suffering bit errors, the monitoring tells us how good the ADC and DAC are, but reveals nothing about the performance of the interface. The interface could be working with 20 dB of noise immunity, or it could be within a whisker of failure. Should the system be marginal such that one bit fails per minute, this will be invisible on an analog monitoring system in the presence of program material and

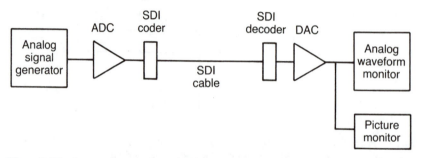

Figure 9.6 Testing waveforms in the analog domain as shown here reveals nothing about the performance margin of the digital interface

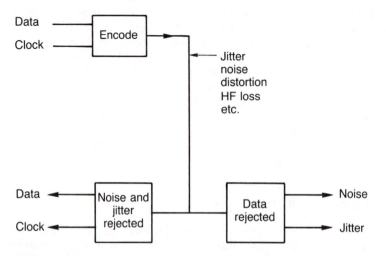

Figure 9.7 The reception process is deliberately designed to reject analog waveform distortions whereas only by measuring these can signal integrity be assessed. Thus the measurement process is diametrically opposed to normal reception and needs special techniques

may just be detectable on colour bars. If the problem is due to a phase-locked loop drifting in an SDI receiver or damp penetration in a cable, it is going to get worse and in the absence of a warning the result will be a sudden failure.

Three distinct testing areas are required in digital video systems. Firstly, on installation, it should be possible to verify that the link is working with an adequate safety margin and that the length of the link is not excessive for the cable type selected. Secondly, it is necessary to test the data integrity of the link to ensure that the BER (bit error rate) is acceptable and remains acceptable when the system is stressed or *margined* beyond the conditions it will experience in service. Thirdly, digital systems can suffer from a problem which has no parallel in the analog domain. This is the protocol error where the data transmission is flawless but the two units concerned cannot understand each other.

Although the SDI signal is digital in that it carries discrete data, it is an analog waveform as far as the cable and receiver equalizer are concerned. Waveform distortions will occur in the cable and noise and jitter will be added. The magnitude of these distortions indicates the likely reliability of the channel. Figure 9.7 shows that a correctly functioning digital receiver is specifically designed to reject the analog waveform distortions by making discrete decisions. By definition, in doing so it denies us knowledge of the signal quality. Thus what is needed is a complementary approach. Instead of rejecting the distortions to obtain discrete data, what is needed is a system which rejects the data in order to measure the magnitude of the distortions.

One approach is to assess the eye pattern generated by the received waveform. After equalization the eye opening should be clearly visible, and the size of the opening should be consistent with the length and type of cable used. If it is not, the noise and/or jitter margin may be inadequate. However, testing the eye pattern on the SDI requires a fast oscilloscope, or even a sampling scope. The scope probe cannot be just hooked on to the cable as the impedance mismatch this causes will have a damaging effect on the signal. Many SDI receivers have a dedicated, buffered, test point for eye pattern monitoring.

Inspection of the eye pattern is acceptable for establishing that the basic installation is sound and has proper signal levels, termination impedance and equalization, but is not very good at detecting infrequent impulsive noise and requires an expensive oscilloscope which has few other purposes in the studio. Contact noise from electrical power installations such as air conditioners is unlikely to be present for long enough to find with an eye pattern display. The technique of signature analysis is better suited to impulsive noise problems.

9.5 Signature analysis

In a digital routing system, reliability is synonymous with data integrity. A data integrity testing system doesn't care what the video waveform is, or indeed that it is a video waveform at all. A data integrity system considers the digital TV field as a block of binary data and simply checks whether that data was received with bit accuracy or not. The message is to forget the pictures and worry about the data.

Signature analysis is a data integrity testing technique which uses large quantities of data and tests down to extremely low bit error rates. As a digital interface which has no bit errors is transparent, signature analysis is a useful way of verifying the transparency of a channel following installation or maintenance. Signature analysis requires a stationary test signal to be applied at the transmitting end of the interface. Stationary means that in the case of component video every frame contains identical data. In the case of composite video the data repeat every four fields (NTSC) or eight fields (PAL). Digitally generated colour bars are suitable but other patterns work equally well. The received data is then processed to generate a value known as a signature which in typical equipment will be a four-digit hexadecimal number. If the transmitted data is always the same, the received signature should always be the same. Any change in the signature indicates that an error has occurred.

The signature generation process divides the incoming bit stream by a polynomial, as was described in Chapter 3. The remainder of the division expresses an entire frame in component or an entire colour sequence in composite as a single word which is the signature. A change of a single bit is enough to change the signature. Any number of bit errors up to the number of bits in the word will guarantee to change the signature, whereas larger numbers of errors are detected with a high degree of probability. In the presence of high error rates the occasional misdetection is irrelevant as the goal of the testing is to determine whether or not remedial action is necessary.

Signature analysis is similar in principle to EDH except that the signature is not transmitted with the data. This avoids complexity at the transmitting end and the need to reserve word positions in the data. As the transmitted data is not a codeword, the remainder will not necessarily be non-zero but could have any value. An error is indicated when the received signature changes, not by its absolute value.

Signature analysis can be used in two ways which are shown in Figure 9.8. In absolute analysis, shown in Figure 9.8(a), the test signal comes from a special generator which displays the signature of the data. The data is fed down the channel under test and then into the signature analyser. The signature displayed on the analyser is compared with the signature on the generator. If the two remain identical for an extended period, then no errors are occurring.

It is also possible to use a relative detection method, shown in Figure 9.8(b). In relative signature analysis, any stationary generator can be used. The correct signature is not known, but if errors occur, the displayed signature will not be stable

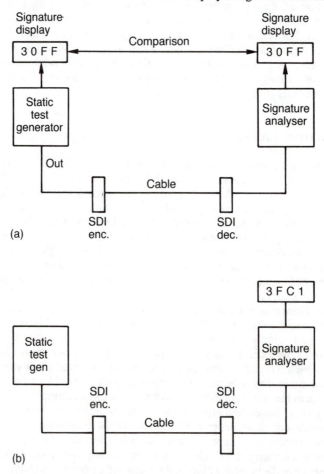

Figure 9.8 In absolute signature analysis (a) the signature at the generator is known and can be compared with the received signature. In relative analysis (b) the signature is unknown but if the generator is known to be static the signature should be constant. Thus changes in the received signature indicate errors

but will change. Thus in relative signature analysis the goal is not that the signature is correct but that it should not change for an extended period. Relative signature analysis cannot detect permanent errors, such as a stuck bit in a parallel interface, as the same signature will always be obtained and so its use is restricted to systems which have no hard faults.

In case of doubt, the test pattern signature can easily be obtained by connecting the generator directly to the signature analyser as well as to the path under test.

Signature analysers can be designed to work on specific parts of the transmission only. If the interface is carrying program material, the signature will vary from frame to frame. However, the ancillary data slots can still be used for signature analysis.

In component systems, the signature analyser may be set to operate on only one selected component. Some machines can be set to operate only on selected bits in the sample, making stuck bits in parallel systems very easy to find.

As signature analysis works in the data domain, it cannot be used to test a channel in which the received data is not necessarily the same as the transmitted data. There are a number of cases in which this could occur. If the digital signal is returned to the analog domain and then converted back to digital, noise in the analog domain will cause data differences. In DVTRs, uncorrectable errors due to dropouts result in concealments which change data values. Thus signature analysis can be used to detect concealment. Note that concealment errors are much less visible than general transmission errors by a factor of about 1000:1, i.e. 1000 concealment errors are about the same level of visibility as one transmission error.

Systems which use compression are not transparent and signature analysis is of no use if a compression codec is included in the test channel. In order to test the channel the generator must be connected *after* the compressor and *before* the decoder.

9.6 Margining

It is a characteristic of digital systems that failure is sudden because deteriorations in the signal are initially rejected until they grow serious enough to corrupt data. Put simply, just because a digital system is working today, there is no guarantee that it will work tomorrow unless its performance margin or 'head height' can be measured. Margining is a long established technique used in the computer industry to prove the reliability of a digital process by testing it under conditions which are more stressful than will be encountered in service. The degree of additional stress which can be applied before failure is a measure of the performance margin. As digital video is only data, margining techniques can be applied to it with great success and if used correctly will give a much needed confidence factor.

There are two ways in which an SDI system can be stressed. The first of these is to use a special signal generator which produces pathological test patterns. These are bit streams which mimic the convolution process of the SDI scrambler and result in channel signals which contain less clock content and lower frequencies than usual. Receivers find it harder to decode pathological signals and so they will result in errors unless the system is in good shape.

A simple alternative to the pathological test signal is the use of a cable simulator which has the same effect as increasing the length of the cable between transmitter and receiver. The Faraday 'cable clone' is the best known of these devices. Figure

9.4 shows the non-linear relationship between cable length and error rate and illustrates the 'crash knee' in the characteristic. Testing with a cable simulator depends upon the existence of the crash knee. Figure 9.9(a) shows how the test is made. The cable under test is temporarily broken and the 'cable clone' is inserted. An error-monitoring system such as EDH or a signature analyser is connected after the receiver under test.

The configuration shown in Figure 9.9(b) is incorrect as it only tests the cable and the transmitter in conjunction with the receiver in the signature analyser. The important receiver is not tested. The cable clone must be installed in the cable and the signature analyser must be connected after the receiver under test.

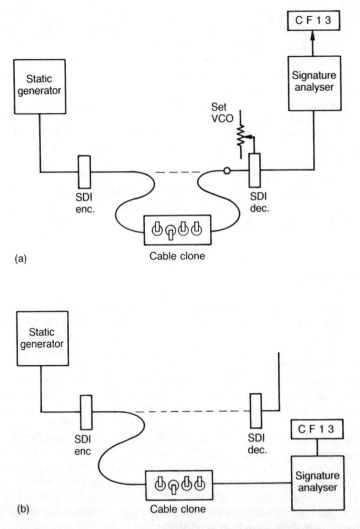

Figure 9.9 Margin testing requires the cable under test to be broken and a 'cable clone' or simulator to be inserted as in (a). The cable length is artificially increased until the crash knee is reached. The configuration in (b) is incorrect as the receiver to be used in service is not being tested. In fact the signature analyser's receiver is being tested: a meaningless exercise

With the cable clone set to bypass, the system should show no errors. If errors are detected the fault should be rectified. Starting with an error-free system the cable length is gradually increased until a rapid increase in error rate indicates that the crash knee has been reached. The additional length of cable needed to reach the crash knee is a direct measure of the performance margin or head height. It will be seen from Figure 9.4 that the error rate changes from negligible to intolerable with an increase in length of only 20–30 metres. Clearly if less than this figure is achieved in the margining test the system is marginal and should not be put into service.

One potential problem area which is frequently overlooked is to ensure that the VCO in the receiving phase-locked loop is correctly centred. If it is not, it will be running with a static phase error and will not sample the received waveform at the centre of the eyes. The sampled bits will be more prone to noise and jitter errors. VCO centring can be checked in a number of ways. If a frequency meter is available, this can be used to display the VCO frequency without input. Another method is to display the control voltage. This should not change significantly when the input is momentarily disconnected. However, the best method is to adjust for minimum error rate in conjunction with a signature analyser (or EDH) and a margining unit or pathological sequence.

Using a cable clone, cable length is added until a slight error rate is caused. The VCO centre frequency is now adjusted one way or the other to see if the error rate can be reduced or even eliminated. If there is a range of error-free adjustment, more cable length should be switched in to make a finer adjustment.

Some SDI receivers run quite hot and the VCO centre frequency changes with temperature. Placing cards on an extender board may change the airflow enough to alter the centre frequency.

9.7 Protocol testing

In digital systems it is possible for problems to occur in which the data transmission is flawless but communication is still not achieved. This can happen where the transmitter and receiver have incompatible protocols. The protocol of a signal includes the nature and positioning of TRS patterns, the location of EDH blocks and embedded audio. A further consideration is that in order correctly to adjust analog-to-digital convertors it is necessary to monitor the actual code values coming from a convertor and compare them with the original analog voltages.

Such problems can only be addressed by a logic analyser. General purpose logic analysers can be used on parallel interfaces, but for best results a dedicated digital video analyser is to be preferred. Figure 9.10 shows the layout of a video analyser. An SDI receiver and/or a parallel receiver drive a decoder which identifies TRS patterns for synchronizing purposes. Incoming words are written into a page of RAM which can hold the contents of a few lines of video. The analyser thus takes a snapshot of interface activity. The point within the frame at which the snapshot is taken is determined by the trigger criteria which are provided to the RAM write logic.

Once the RAM is written in real time with the digital video snapshot, the data are frozen and can be inspected using the associated PC and its display. Signals can be displayed graphically as they would appear on conventional analog waveform monitors or vectorscopes, but can also be displayed as tabular data so that the exact binary word values and positions can be established. Any departures from correct protocol can be detected by comparing the data with the relevant standards.

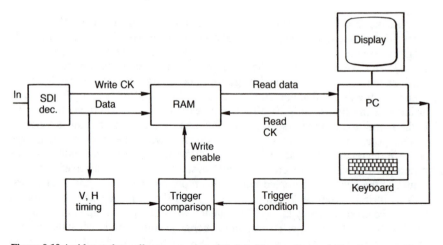

Figure 9.10 A video analyser allows a snapshot of digital video interface activity to be captured in RAM. This data can then be inspected at leisure using a small computer

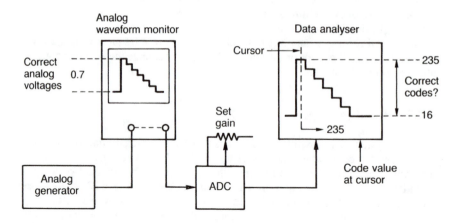

Figure 9.11 Using a video analyser to set the gain of an ADC. The ADC output data is compared with standard values on a known part of the analog waveform

Figure 9.11 shows how a video analyser is used to set up a component ADC. Convertor set-up is critical as digital systems keep the data the same all down the line, so it has to be right from the outset. A precise analog test pattern generator is used to produce colour bars. The signal is looped through an analog waveform monitor to the ADC where it should be terminated with the terminator which will be used in practice. This is necessary because terminator tolerance is such that changing a terminator can change the analog level by several digital code values. The generator level is adjusted until the waveform monitor displays exactly the correct amplitude in, say, the white bar.

The analyser then captures the ADC output and the black bar is located in waveform mode with the cursor. The actual data values at the cursor position are displayed, and these can be compared with the ideal so that any offset in the convertor can be revealed and adjusted out. By selecting the white bar the luminance convertor gain can be adjusted. Some analysers allow timed retriggering so that a new snapshot is taken periodically. This mode is useful when making dynamic gain adjustments. The colour difference signal gains are set up in a similar manner.

The analog Y/C timing should be checked with a suitable signal such as bowtie before returning to colour bars to check the relative timings of the green/magenta transitions in the three digital components. This is not as easy as it sounds as the colour difference sample spacing is twice that of luminance and the sample values need to be interpreted carefully to find the transition.

Index